Motorbooks International
POWERPRO SERIES

HOW TO BUILD & MODIFY
CYLINDER HEADS, CAMSHAFTS & VALVETRAINS

Ben Watson

First published in 1993 by Motorbooks International Publishers & Wholesalers, PO Box 2, 729 Prospect Avenue, Osceola, WI 54020 USA

© Ben Watson, 1993

All rights reserved. With the exception of quoting brief passages for the purposes of review no part of this publication may be reproduced without prior written permission from the Publisher

Motorbooks International is a certified trademark, registered with the United States Patent Office

The information in this book is true and complete to the best of our knowledge. All recommendations are made without any guarantee on the part of the author or Publisher, who also disclaim any liability incurred in connection with the use of this data or specific details

We recognize that some words, model names and designations, for example, mentioned herein are the property of the trademark holder. We use them for identification purposes only. This is not an official publication

Motorbooks International books are also available at discounts in bulk quantity for industrial or sales-promotional use. For details write to Special Sales Manager at the Publisher's address

Library of Congress Cataloging-in-Publication Data

Watson, Ben.
   How to build and modify cylinder heads, camshafts & valvetrains/ Ben Watson.
      p. cm. —(Motorbooks International powerpro series)
   Includes index.
   ISBN 0-87938-790-4
   1. Automobiles—Motors—Cylinder heads—Maintenance and repair. 2. Automobiles—Motors—Camshafts—Maintenance and repair. 3. Automobiles—Motors—Valves—Maintenance and repair. 4. Automobiles—Motors—Modification. I. Title. II. Series.
TL214.C93W38   1993
629.25'2—dc20   93-31731

On the front cover: A custom Chevrolet small-block head and trick camshaft are among the components developed by Myron Cottrell's Tuned Port Induction Specialties (TPIS) in Chaska, Minnesota. *Eric Miller*

Printed in the United States of America

Contents

1   **Where Does the Power Come From?** 4
2   **What Does the Cylinder Head Do?** 15
3   **Tools** 19
4   **Fasteners** 28
5   **Choosing Your Heads** 35
6   **Cylinder Head Reconditioning** 38
7   **Airflow, Camshafts and Horsepower** 48
8   **Warped and Cracked Heads** 62
9   **Cylinder Head Measurement** 68
10   **Valves, Valve Guides, Seats, & Springs** 85
11   **Blueprinting** 93
12   **Major Valve Refitting** 118
13   **Engine Reassembly, Camshaft and Valve Adjustment** 126
14   **Other Performance Modifications** 134
15   **Tuning the Engine** 143
16   **The Law and Engine Modifications** 158
      Drill Sizes for Taps 159
      Index 160

Introduction

In the introduction to my book *How to Rebuild Your Engine* I recounted the story of the first engine I ever rebuilt. Less than three years later Bill and I found ourselves building our first racing engine. Actually, calling it a racing engine really only refers to its intended destiny. We bought an old Ford 289 block with 12:1 compression pistons, added a slightly better than stock cam and used the original heads. Then we went racing.

The racing club we belonged to adhered closely to the SCCA (Sports Car Club of America) rules of that time. Now I do not want to imply that we operated on a low budget but we only had one Nomex fire suit for two drivers. Rules required that each driver get a minimum number of practice laps before acquiring their license to race.

To this day, twenty years later, former members and spectators still talk about the team that would strip in their hot-pits, not only changing drivers, but the drivers sharing clothes. (The reader should note that we only shared the Nomex suit, we each had our own Nomex undies.)

In addition to the shared firesuit, we took other cost-cutting-measures. The sway bar was the torsion bar from a 1970 police package Plymouth Grand Fury with angle iron and helm joints welded on, the carburetor overflow tray was a cooking sheet with a hole cut in the middle and held on with picture hanging wire. The sealed oil breather cap was a crutch tip while the oil catch can was the brake fluid can we had used to fill the master cylinder after overhauling the brakes. The roll bar was a bolt-in model purchased through mail order. When it arrived we found two problems with it. First, the bar mounted to the floorboards. My guess was the floorboards of this old Mustang had rusted out in the late 1960s. The second problem was that the roll bar was for a notchback Mustang, our race car was a fastback. We cut panels from an old MG hood and pop-riveted them over the top and bottom of the floorboards to hide the rust-out. We used an old camp stove to melt lead diving weights into wedges to fit under the feet of the notchback roll bar. When we were questioned about these things during tech inspection we told the inspectors that we were concerned about mounting the roll bar to the floor and added the extra panels and lead to "spread out the load." Bill suggested in private that if the car did not roll over the roll bar would hold fine . . .just long enough for you to feel as though you were safe, then punch through. He also pointed out that due to my propensity not to keep the car between the edges of the pavement, when the roll bar did punch through the car would disappear in a cloud of body-putty dust.

In spite of the financial limitations of those early days, racing became a good training ground for me. I soon found that the "racing engine" we had built had less power potential than the stock engine we had started with a season earlier. What was the chief reason for the limitation? Why could we not get the potential power out of the engine? The cylinder head!

Where Does the Power Come From? 1

One of the most powerful forces on earth is something we cannot see, something we cannot hold in our hand, but something that has the power to flatten entire cities. Humans have been harnessing this power to propel vehicles for thousands of years. Is this some recently rediscovered secret of ancient Egypt we are talking about? No, it is the power of air. Ask any tornado survivor.

Individual air molecules interweave and layer themselves like bricks in a wall. Cooler air, like the air in your freezer, for instance, is relatively calm and takes up little space. Warmer air, on the other hand, displays some interesting characteristics. For example, the air in your oven is air "with an attitude." Each molecule is moving quickly to protect its territory. Not only are the molecules trying to protect their own territory, but they are also trying to take part of their neighbors' territory. The hotter these active molecules are, the more space they take up.

An internal-combustion gasoline engine creates power through a series of precisely coordinated events. First, a proper ratio of fuel and air must exist in the air intake system. Second, this air-fuel mixture is guided into the combustion chamber through the intake manifold, cylinder head intake ports, and intake valve. The air is then compressed in the cylinder by the upward movement of the piston. As the air is compressed, it gets hotter and becomes more active. As the piston approaches the top of its travel, the ignition system ignites the air-fuel charge and the air becomes increasingly active, driving the piston downward as

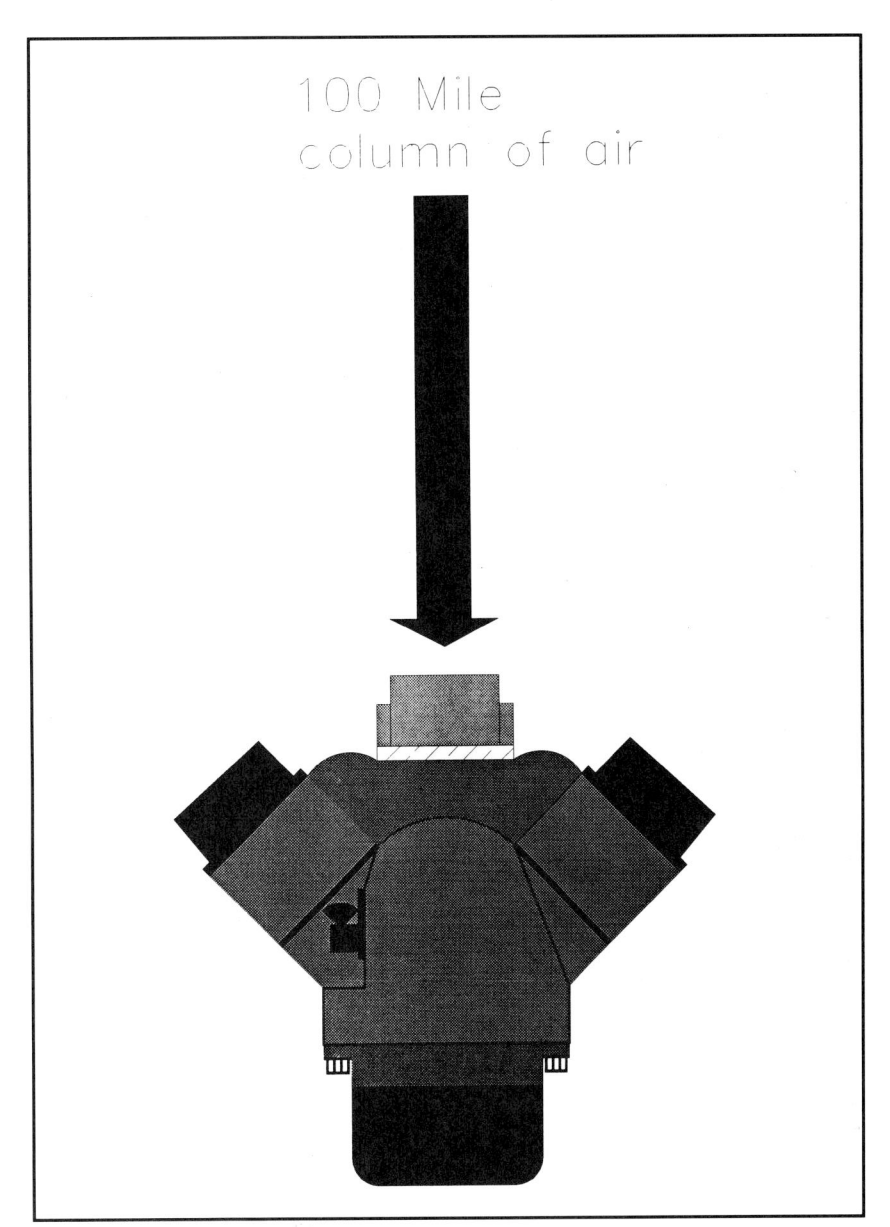

The power from an internal combustion engine is dependent on a gaseous mixture the surrounds the engine and all of us. That mixture is commonly called air. As a piston starts its intake stroke a low pressure area is created in the cylinder. A 100-mile-tall column of air forces a mixture of 21 percent oxygen, 78 percent nitrogen, and 1 percent miscellaneous gases into the combustion chamber. In the combustion chamber, fuel and heat are added to the oxygen element of the air. Ignition occurs and rapidly expanding gases force the piston down on the power stroke. Power in a gasoline, spark ignition engine is a function of how much air can be forced into the combustion chamber.

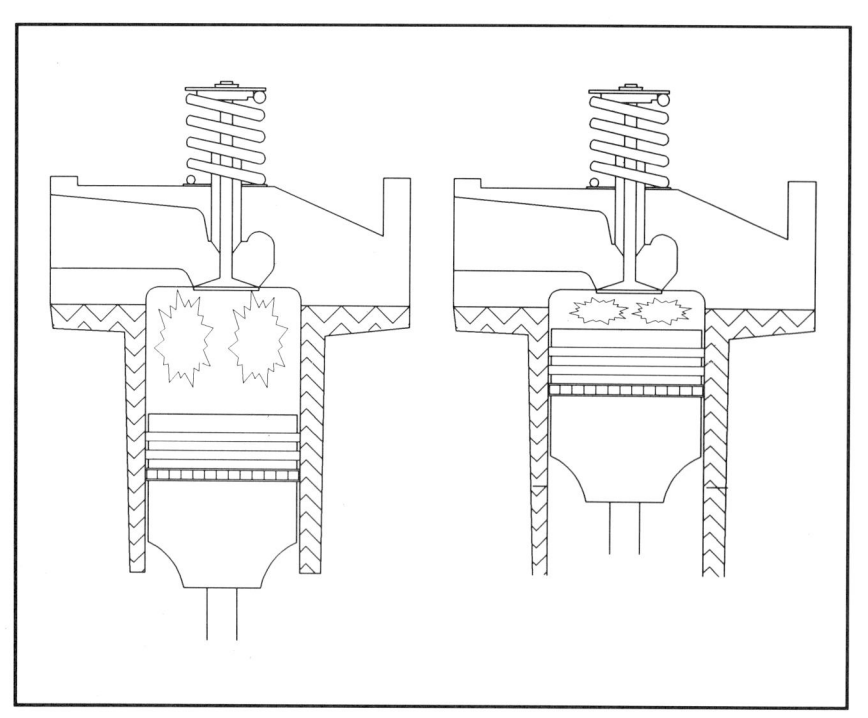

As the piston begin its journey up the cylinder, the gases (air) in the combustion chamber are rather calm, they are rather cool. As the piston approaches the top of the compression stroke the rather calm, docile molecules begin to be crowded for space, they get an attitude, they get hot. This heat helps in igniting the air/fuel mixture.

This "flathead" technology was being phased out as state of the art when I was born. In those days power came from size. This inefficient design led to incredibly large engines. A great example of this was the 15,000cc (915 cubic inch) 4-cylinder engine made by Fiat in the 1920s.

the molecules fight for territory. As the piston displaces downward, there is less of a fight for territory and the molecules begin to become less active. The gases that resulted from combustion begin to cool. Soon, the exhaust valve opens and the crankshaft swings past bottom dead center. Momentum, as well as the highly active air in other cylinders, forces the piston up, pushing the gases out through the exhaust valve, through the exhaust ports of the cylinder head, and out through the exhaust system.

This book focuses on what occurs from the time the air moving through the engine leaves the intake manifold, until the time it enters the exhaust system. Power in an internal-combustion, spark ignition, gasoline engine comes from the rapid expansion of hot air. The trick to getting power from an engine is to get the air into the engine and get the burned exhaust gases out of the engine as quickly as possible. Of all performance modifications, the ones that contribute most to releasing the engine's potential power are the modifications that can be made to the cylinder head and valvetrain. There is a tendency for the neophyte to think that more power is best achieved by putting more fuel into the engine. Anyone who has ever experienced a sunken carburetor float or leaking fuel injector knows that fuel alone does not increase power. Of course, the same is true of air. Air without enough fuel mixed with it will not be heated properly, wasting the effort to pull it into the engine. Fuel is easy to get into the engine; air is difficult.

Back in the days of the 409, the Beach Boys, and Ed Sullivan, the manufacturers got more power from their engines by gulping large volumes of air into even larger cylinder displacements. This would be analogous to a Sumo wrestler approach—power through bulk. Today, power is achieved by coercing as much air as possible into as small a combustion chamber as possible. Power through efficiency. This would be equivalent to the Greco-Roman wrestler approach.

Early engine performance amounted to making the engines larger. During the 1920s Fiat built a 15-liter engine, that is, 915 cubic inches (ci). As incredible as that might sound, it is doubly incredible when you find out it was a four-cylinder engine. The trend for bulk rather than efficiency dominated the industry through the late 1960s. During this heyday of the land rocket, Detroit turned out very inefficient (by today's standards) gas-guzzling monsters. By comparison, one of the latest, more innovative engines out of Detroit is the General Motors Corporation Quad 4. This 150ci engine can crank out 154 horsepower (hp). If the popular performance engine of the 1960s, the 396, had been as efficient as the Quad 4, it would have cranked out 402hp instead of the more docile 300 to 350hp. Although several technological improvements have been made permitting increased horsepower per cubic inch, more efficient cylinder head design—a tip learned from the 1960s and 1970s—has been a major contributing factor. Better airflow and improved combustion chamber design are not the least of these factors, and are areas that will be addressed in this book.

Years ago, as a friend and I were building our first racing engine, our neighbor in the mini-warehouse complex, Ron, was building a Datsun 510. Since we were building a small-block Ford V-8, and since in those days Datsuns were still as rare as a Chevy lover at a Mustang convention, we thought that Ron was either weird or out of his mind. During the past twenty years or so, however, I have come to realize that a lot of exciting performance work *can* be done on four- and six-cylinder engines. Imagine a Volkswagen Scirocco with a basically stock engine that could break off the transmission mount bolts with engine torque. Or a SAAB with a mostly stock engine that allows you to look eyeball to eyeball with a Corvette owner during that grudge race. Modifications to small-engine cylinder heads often can offer more satisfaction than reworking some huge piece of iron. A 409 (6.7 liter) operating with a volumetric efficiency of 60 percent would produce no more power than a 4.0 liter operating at 100 percent volumetric efficiency. This book is the story on how to get more out of less.

Let's begin by tracing the history of cylinder head and valvetrain design all the way back to Nikolas Otto in the late 1800s. In 1876, Nikolas August Otto and Eugene Langen obtained a US patent for two-cycle and four-cycle gas engines. These early engines were not powered by gaso-

In the flathead design the valves were in the block and moved parallel to the pistons. This meant the air destined for the combustion must turn a 180 degree corner before entering the combustion chamber. The result was very poor volumetric efficiency.

line, but by a variety of gases. The engine featured a cylinder head and valve design that later would be known as the L-head design.

Manufacturers have created a number of cylinder head designs over the years. In this chapter we will trace the history of various types, and discuss both their advantages and disadvantages. This information will assist later in our discussion of redesigning and replacing cylinder heads for performance.

The scope of this book also will include replacement and reworking of cylinder heads for overhead valve and overhead cam engines. While modifications to these two types of heads are similar in most respects, portions of each operation do differ somewhat.

Counterflow and Cross-Flow Heads

In terms of airflow through the cylinder head, there are two basic styles of heads: counterflow and cross-flow. On counterflow heads, intake air and exhaust gases pass in and out through the same side of the cylinder head. This limits the breathing ability of the engine and therefore the potential space in the combustion chamber for fresh, combustible air. Assembly-line production makes this design common on many inline engines. The Ford 4.9 six-cylinder truck engine is a familiar example of the counterflow head design. Many import engines, particularly those destined for transverse installation, use this design as well.

For the performance-oriented enthusiast, the counterflow design makes turbocharging easy. Since the exhaust and intake manifolds are on the same side of the head, there is less plumbing and rerouting of the exhaust system to power the turbine side impeller of the turbocharger.

The major disadvantage that makes this design increasingly rare, however, is that the intake sits immediately above the exhaust. As a result, heat can transfer from the exhaust to the intake. In the days of smog-belching gas guzzlers, this was no dis-

7

Counterflow head

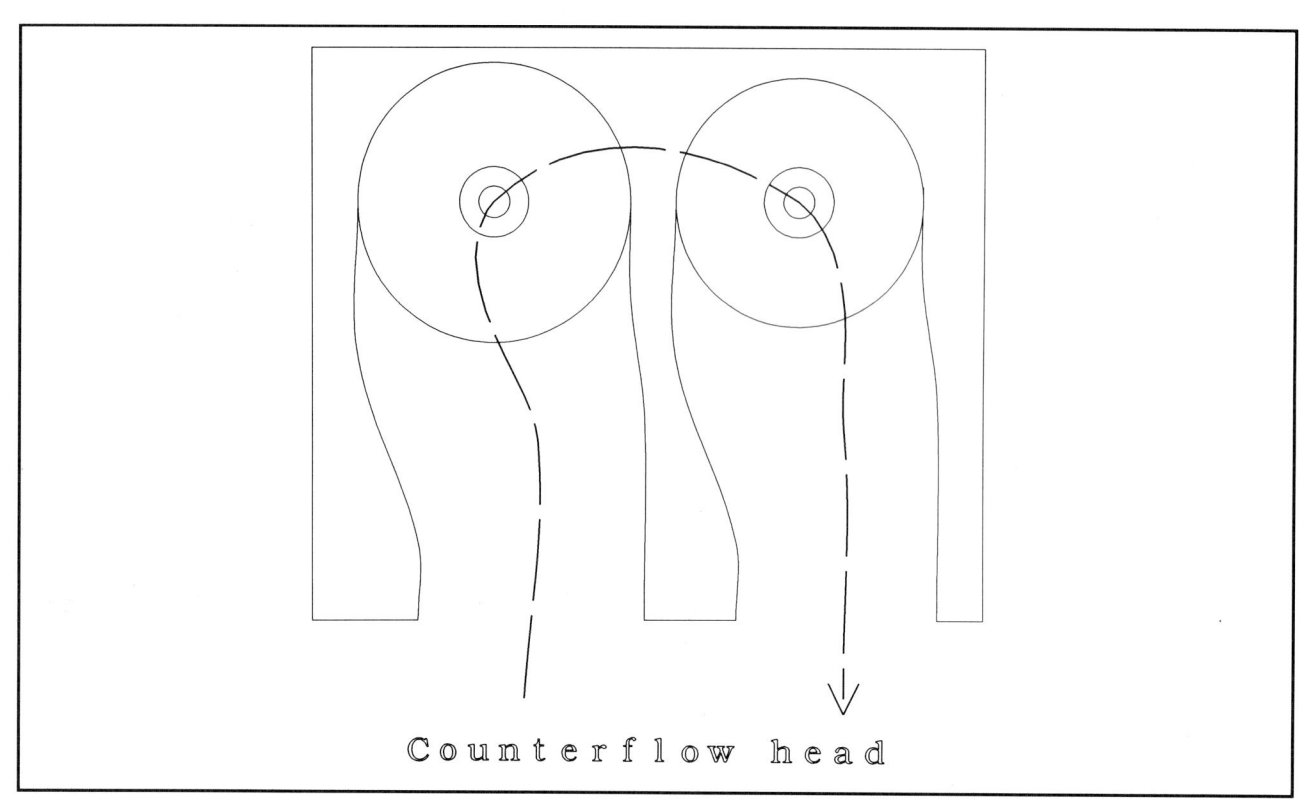

The relative location of the intake and exhaust system can go a long way toward dictating how efficiently air can flow into the cylinder and exhaust gases out of the cylinder. When the intake and exhaust are located on the same side of the engine the air must flow in from one side of the engine and after combustion reverse direction and exit as exhaust gases through the same side of the engine. Think of it this way, how much more effortless would mowing the lawn be if you never had to change the direction of the lawn mower.

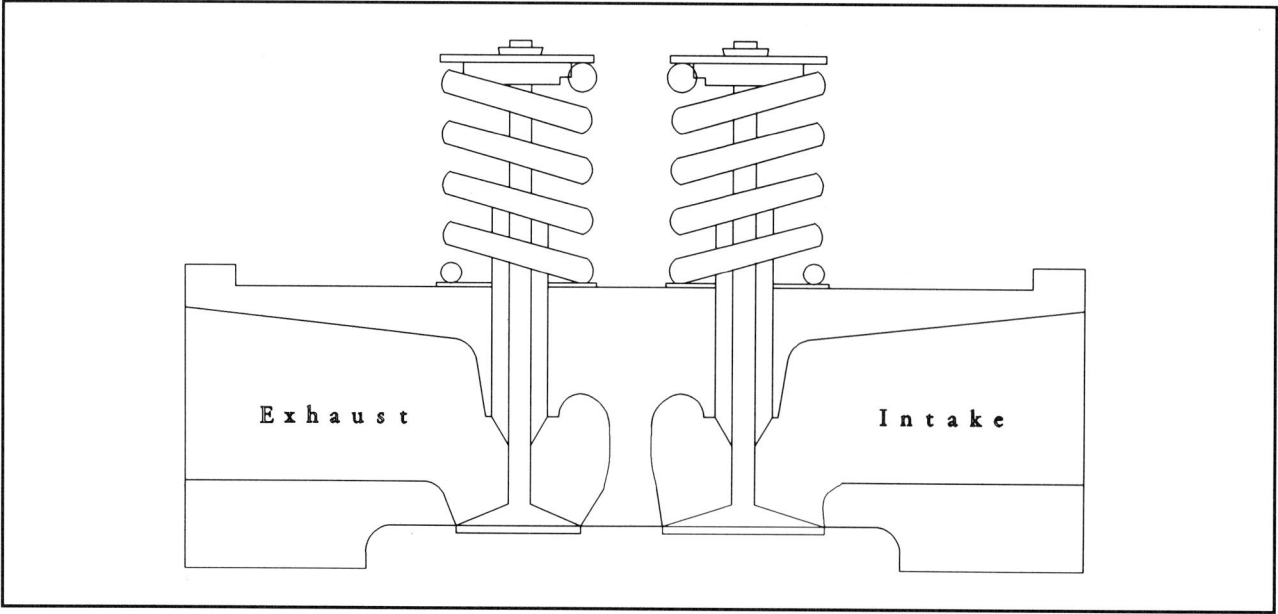

In this cross-flow cylinder head design the inertial mass of the gas (in the form of air on intake and in the form of exhaust gases on exhaust) never has to reverse direction. The result is a better flow of gases through the head. Additionally, during the brief time when both the intake and exhaust valves are open expulsion of the exhaust gases helps to draw in the intake air.

advantage. As air-fuel ratios became leaner, more toward the point of the chemically perfect ratio known as the stoichiometric point, manufacturers using this design began to experience inherent driveability problems. As the heat transferred from the exhaust to the intake, fuel would begin to boil in the bowl of the carburetor. In some cases, this problem was overcome with the use of throttle body or port fuel injection.

A more popular cylinder head design for performance-oriented people is the cross-flow head. In a cross-flow design, the intake air-fuel mixture enters through one side of the head and exits through exhaust ports on the opposite side. The advantage of this design is obvious: the momentum resulting from the straight-line gas flow helps to "supercharge" the intake side while scavenging the cylinder on the exhaust side. This effect is especially notable during valve overlap. Briefly, toward the end of each exhaust stroke and the beginning of each intake stroke, both the intake and exhaust valves are open. The time during which both valves are open is known as valve overlap. For normally aspirated engines (no turbocharger or supercharger), the cross-flow design has distinct advantages over the counterflow design.

Although it may not need to be said, modern V-8 engines all use the cross-flow design. Since the 1950s, they have been the most popular engines to modify for high performance.

Since air is invisible, we tend not to think of it as a real substance with physical properties and specific flow characteristics. Therefore, in an attempt to speak in more concrete terms, throughout the book an analogy will be made to something more tangible and thus more easily understood—water.

In a sense, air flowing into a combustion chamber is a lot like the surf beating against the shore. When the intake valve opens, the atmosphere pushes air into the cylinder. Imagine a small cove, similar to the beach visited by Burt Lancaster and Deborah Kerr in the film *From Here to Eternity*. Outside the cove is the vast expanse of the Pacific Ocean. The waters of the Pacific are always trying to fill the cove. However, since water seeks its own level and since we have no way of emptying the cove, the only water that will enter through wave action from the Pacific is the amount that will fill the lowered level of the cove.

At the beach end of the cove, let's place a large, flat panel driven by a huge steam-powered ram. Between waves, the ram pushes the majority of the water in the cove back out into the ocean. Notice that not all the water in the cove is pushed into the ocean. This is because it is impossible to completely seal the panel against the floor of the cove. In the same way, it is impossible for the piston to seal against the cylinder walls. The piston has to force exhaust gases into the same atmosphere that is trying to force air into the combustion chamber.

We begin the process by forcing the water out with the steam-powered panel. Much of the water has been removed from the cove, but there is still a little left behind. This will reduce the amount of new water that can enter; in a cylinder, this would reduce the amount of fresh air available to support combustion. Now, the race begins: Will the water at the mouth of the cove be able to force its way through the mouth of the cove at the same rate as the withdrawal of the panel? There is little doubt that it will, when the speed of the withdrawing panel is relatively slow. As the speed of the panel increases, however, the speed of the water rushing through the mouth of the cove does not increase. As a result, the faster the panel moves, the less likely the cove is to fill completely. Between the water left behind when the panel attempts to clear the water and the tardiness of the water to enter through the mouth of the cove, the amount of fresh water entering the cove during each cycle is far less than the total capacity of the cove. In the internal-combustion engine, the percentage relationship between the total water capacity of the cove and the amount of new water taken in each time is called volumetric efficiency.

The cylinder head acts like the inlet to the cove. As does the inlet, the head's intake and exhaust ports limit the speed at which the ocean of air can push the new air into the combustion chamber and accept the exhaust gases. At different points through this book, we will be adding outlets to the cove, reefs, and rocks.

Boyle's Law

This principle is essential to the operation of the normally aspirated four-stroke internal-combustion engine. Boyle's law states that the pressure of a stationary fluid will apply an equal force in all directions. When the piston drops down the cylinder, it creates a void. If there is no cylinder head, the void is immediately filled to a pressure that is 100 percent of atmospheric. If there is a cylinder head and no intake valve, or if the intake valve remains closed, a high vacuum is created. If the intake valve opens slightly, the cylinder fills to a partial vacuum, a small percentage of atmospheric pressure, and there is a small amount of power created by the engine.

There are restrictions to airflow in two major areas. The first is in the throttle plates. Whether in the throttle bores of the carburetor or the throttle bores of the fuel injection throttle body, the driver uses these to control the power output of the engine. The

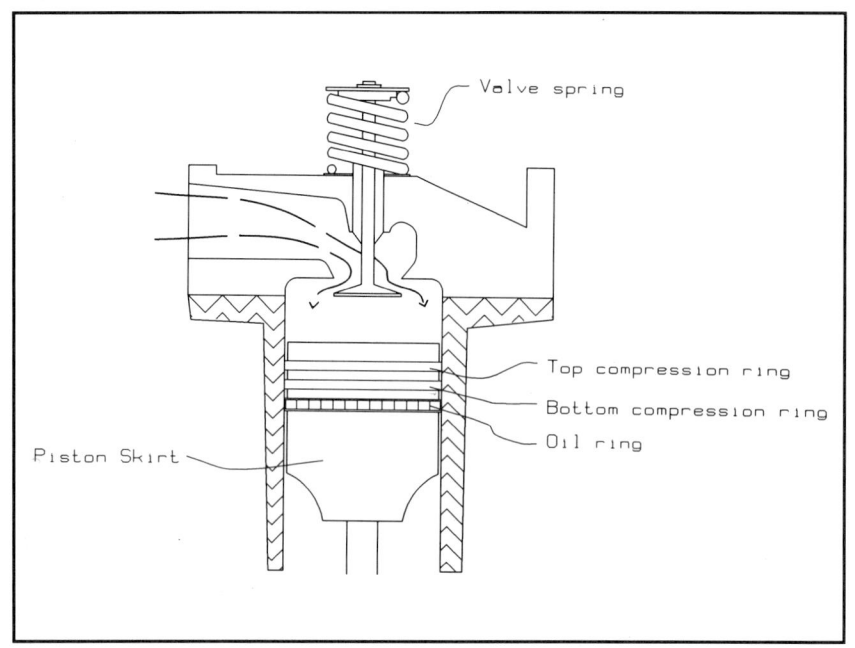

"Nature abhors a vacuum." As the piston starts its downward journey in the cylinder, a low pressure area is created. Boyle's law demands that the pressure in the cylinder and the pressure in the atmosphere try to equalize. Air rushes into the cylinder. The pressure in each of the cylinders becomes equal to atmospheric at the bottom of the intake stroke, which is much harder than it seems, then we say the engine has a 100 percent volumetric efficiency. Most normally aspirated performance engines have 100 percent volumetric efficiency at some engine speed, but none have it at all engine speeds.

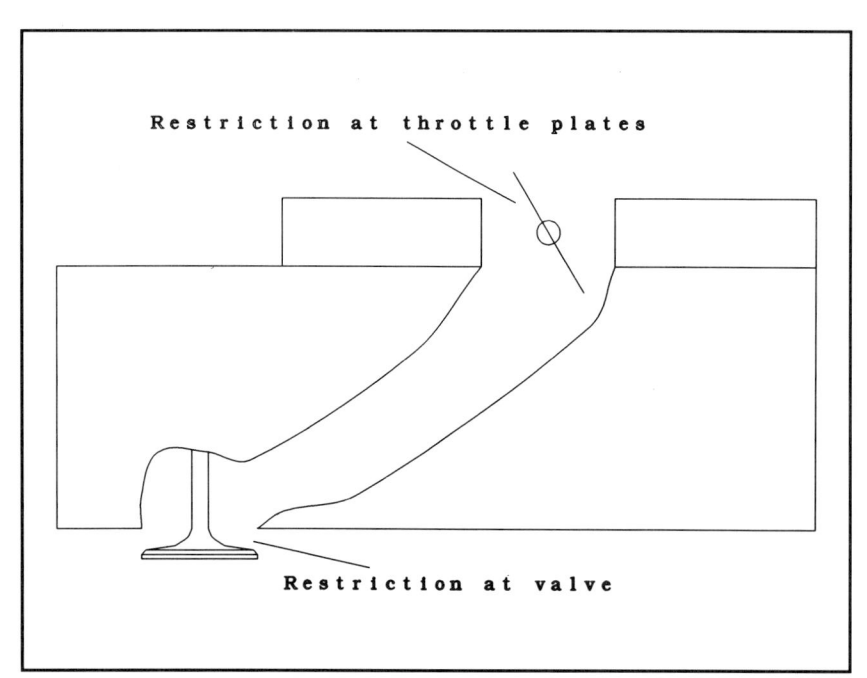

Several restrictions prevent 100 percent volumetric efficiency. There is the restriction of the air cleaner and air filter. There is the restriction of the throttle plates and the intake manifold. In the cylinder head there are the restrictions of the casting pits and ridges of the intake and exhaust ports and valve pocket. The valve stem and valve also restrict air flow.

second area of restriction is the intake valve. If the throttle plates are used to control the power and speed of the engine, the intake valve ideally would allow the cylinder to fill to 100 percent of atmospheric pressure (or maybe more?).

A 302ci Ford engine that has been bored 0.030in oversize displaces a volume of 0.022 cubic feet per minute (cfm), or 38.25ci, each time the piston drops. If the engine is running at 5000rpm, each cylinder could conceivably draw in 55.3cfm of air. At 5000rpm, the engine has the potential of drawing in 443cfm of air. Anyone who has ever thumbed through a performance catalog knows that there are plenty of carburetors and fuel injection systems that have potential airflow rates greater than 443cfm. The trick to getting the air into the cylinder is not routing it through the carburetor bore, not even routing it through the intake or exhaust manifold. The trick is to get *through* the cylinder head and *around* the intake valve.

An equally important task is to exhaust the burned and partially burned combustion gases. If the cylinder is still partially filled with the burned gases when the intake valve opens, the cylinder cannot fill completely with fresh air. The piston rushing up the cylinder helps evacuate these residual gases, but the restriction formed by the cylinder head's exhaust valve and ports slows the movement of these gases. This restriction causes the piston to pressurize the gases in the combustion chamber instead of pushing the gases out. The gases that are simply compressed expand during the intake stroke, limiting the amount of fresh air to fill the cylinder.

Actually, the exhaust gas volume will exceed the intake volume. In a typical situation the temperature intake air will be less than 120 degrees Fahrenheit, while the exhaust gases will be well in excess of 600 degrees. At atmospheric pressure, thermal expansion will cause the gases at 600 degrees to occupy a much greater volume than the same gases at 120 degrees. Therefore, the exhaust system must be able to flow more gas volume than the intake. Again, anyone who has perused a performance catalog knows that exhaust systems are available that can easily flow 443cfm.

So, it is fairly obvious that the most significant restriction to airflow occurs in the cylinder head. Keep in mind, though, that these figures represent the extreme. Under normal, legal driving conditions, gas flow rates never come close to these numbers. A person deciding to modify the cylinder head on his or her engine might be disappointed if the same results are not achieved. In addition, the typical commuter will derive little benefit from additional power obtained, and he or she will notice little difference in fuel economy. In reality, both have been improved. When entering freeway traffic, the commuter will be able to accelerate faster; when fuel economy is monitored over a period of several weeks, head modifications will have improved mileage.

Bernoulli's Law

Above the cylinder head there is a column of air nearly a hundred miles tall. So says Bernoulli's Law. This column creates a pressure on the throttle plates of 14.7 pounds per square inch (psi). The dropping piston creates a rapidly dropping pressure in the cylinder. Let us look at a hypothetical example. If the intake valve remains closed while the piston drops, and if the compression ratio is 10:1, and if the combustion chamber is filled to atmospheric pressure when the piston is at the top of the stroke, and if the displacement of the cylinder is 37.75ci, then when the

The biggest advantage of the "V" engine design is cross-flow design with the exhaust ports lower than the intake ports. This old 302 Ford engine is bored 0.030 over and has a theoretical air flow of 443cfm at 5000rpm. The heads however were rescued from the wrecking yard in the early 1970s. The castings are for an early 1960s 221 V-8. These heads will probably limit the air flow to well less than the theoretical 443cfm.

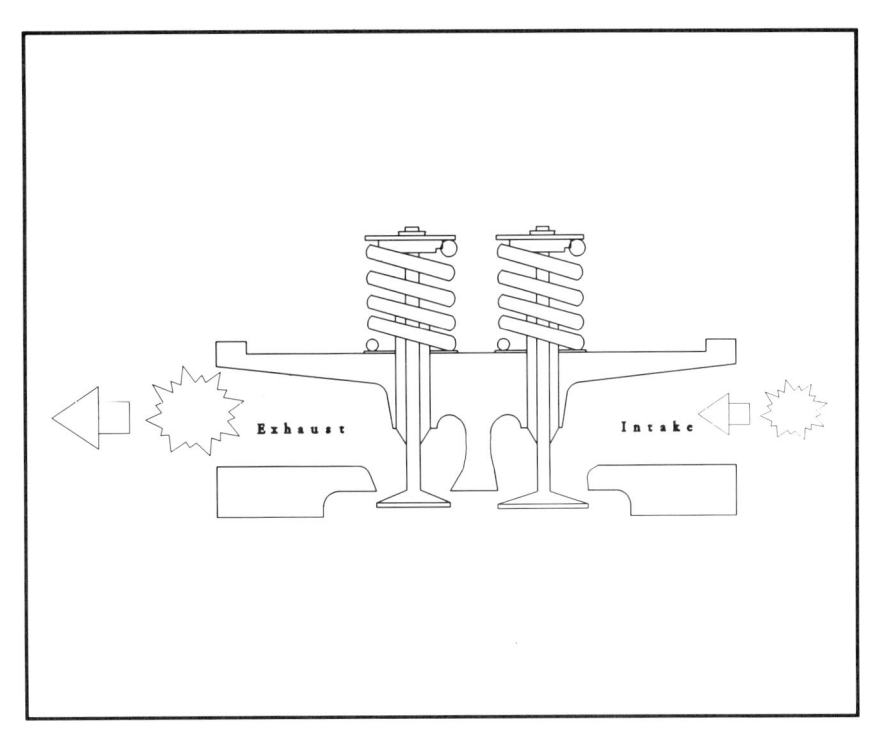

The gases exiting the cylinder head are much greater in volume than those entering the cylinder. The mass of the gases has only been changed with respect to how much the fuel has added. The weight of the air in the combustion process is 14.7 times greater than the weight of the fuel, therefore the mass of the gases increases only slightly during combustion. The volume of the air entering the engine increases quite dramatically with respect to the amount of temperature increase during combustion.

A force of 14.7psi presses in on all objects at sea level, people and engines. This pressure forces air into the cylinder when the intake valve opens and the piston drops down. This air must make its way through the twists and turns of the intake manifold and through the twists and turns of the cylinder head ports.

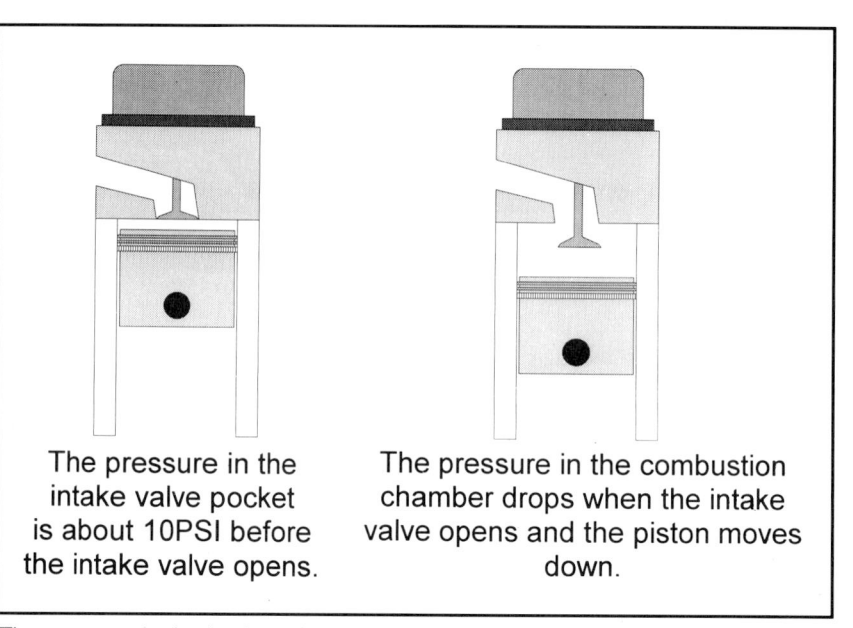

The pressure in the intake valve pocket is about 10PSI before the intake valve opens.

The pressure in the combustion chamber drops when the intake valve opens and the piston moves down.

piston reaches the bottom of the stroke, the pressure in the cylinder will be 1.33psi. If the intake valve is open while the piston is traveling downward, then a race will be on between the downward movement of the piston and the air flowing through the intake ports and valves of the cylinder

The pressure in the intake valve pocket is slightly higher than the average pressure in the manifold. When the intake valve opens and the piston drops

head to fill the partially vacuous cylinder. To fill the cylinder to atmospheric pressure, 37.75ci of air at atmospheric pressure must flow into the cylinder. If the cylinder can fill to atmospheric pressure at the same rate the piston is dropping, then we can say that the cylinder has a 100 percent volumetric efficiency at that rpm and piston speed. If only 30.20ci of air can flow at that piston speed, then the cylinder has an 80 percent volumetric efficiency.

Imagine for a moment the high-performance engine of the twenty-fifth century. The cylinder head is replaced by a force field. During the intake stroke the force field is switched off, exposing the cylinder bore to direct filling from atmospheric pressure. During the compression and power stroke, the force is reactivated to seal the cylinder. The piston speed would have to be incredibly high before the cylinder could *not* fill to full atmospheric pressure. The same is true of exhaust gases; disabling the force field would make it possible to vent the contents of the cylinder

down the pressure in the valve pocket and the cylinder tries to equalize, the cylinder fills with air.

to the atmosphere through the entire cross-section of the cylinder bore.

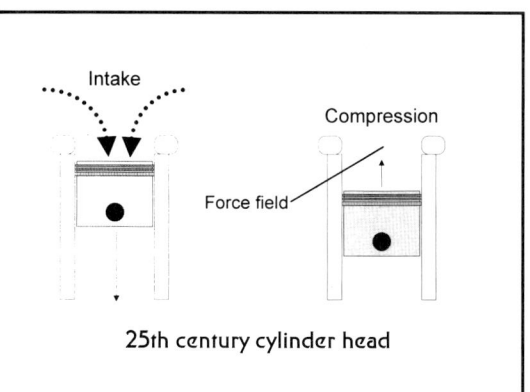

Unfortunately, in the crude realities of the twentieth century we do not have cylinder force fields. We must use cylinder heads, which instead of opening the combustion chamber to the entire atmosphere can only supply passageways for the movement of the air-fuel mixture and the exhaust gases. These passageways, when improperly designed or modified, can adversely affect the flow of gases.

Thermodynamics

Thermodynamics is a science designed by short men with plastic pocket liners and a preference for slide rules rather than calculators. Beyond the ability of us mere mortals to fully comprehend, thermodynamics deals with the expansion and contraction of gases as heat and pressure are changed. *Webster's New World Dictionary* defines it as "the branch of physics dealing with the reversible transformation of heat into other forms of energy,

25th century cylinder head

If I could talk to Scotty or Gordy LaForge I would ask them to design a cylinder head that consists of a force field. During the intake stroke the force field would be switched off. As the piston drops, the cylinder fills immediately and completely with air. During the compression and power stroke the force field would then seal the cylinder. The force field would be dropped for the exhaust stroke.

especially mechanical energy, and with the laws governing such conversions of energy."

As the air and fuel are drawn into the combustion chamber they are compressed by the rising piston. The pressure in the combustion chamber begins to rise. As the pressure rises, the temperature also rises. This rise in temperature is predictable for the air-fuel charge and serves to warm the charge, making it more easily combustible. When the spark plug sparks, the air-fuel charge ignites in the rapidly rising temperature, causing the air and the combustion gases in the cylinder to expand. The expanding gases are driven into the combustion chamber, converting the heat energy into mechanical energy. The piston is forced downward by the expansion of the combustion gases, the crankshaft rotates, and the car is propelled down the road.

The Otto Cycle Engine

As mentioned, the modern automotive gasoline engine is a

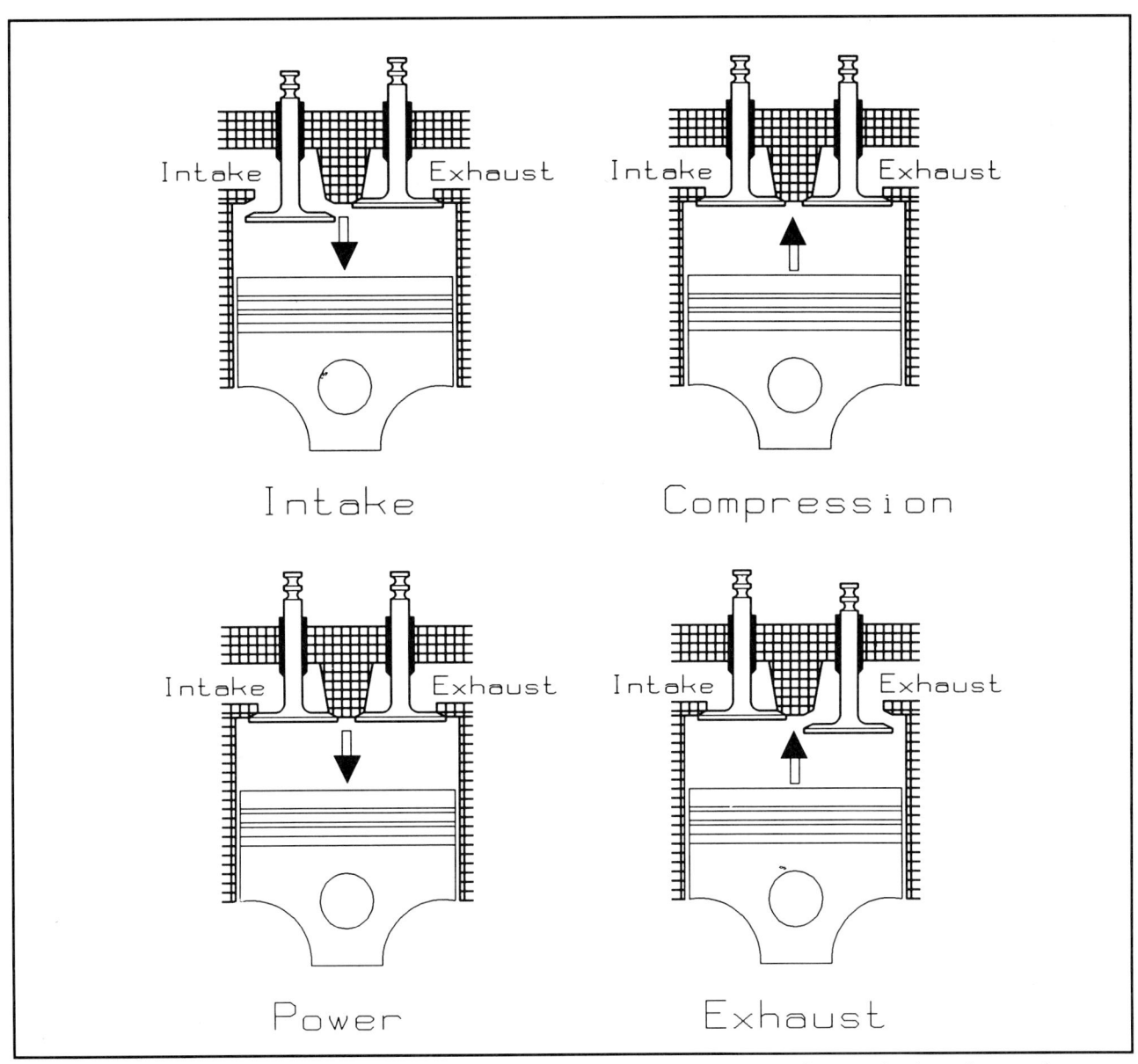

The Otto cycle engine has four strokes. During the intake stroke the intake valve is open as the piston moves down. Air is pulled into the cylinder. During the compression stroke the intake and exhaust valve are both closed. The pressure in the cylinder rises, and as the pressure in the cylinder rises, the temperature rises. During the power stroke both valves are closed, the heated expanding gases force the piston down. During the exhaust stroke the exhaust valve opens; as the piston rises the spent gases are forced out of the combustion chamber by the rising piston.

four-stroke engine. The four strokes for which it was named include the intake, compression, power, and exhaust. While these terms have long been associated with the operation of the internal-combustion four-stroke engine, I prefer the far more descriptive terms used by a Canadian associate. I have modified his terms a little to make them more printable: suck, squeeze, boom, and belch.

Let's begin with the intake, or suck. At the beginning of the intake stroke, the intake valve opens. Synchronized to the opening of the intake valve, the piston starts one of its downward strokes. This downward movement of the piston creates a low-pressure area in the cylinder. Air and fuel rush into the cylinder through the open intake valve. At the bottom of the piston travel during the intake stroke, the cylinder is charged with air and fuel. Rarely does the cylinder fill to atmospheric pressure with the air-fuel charge. As mentioned earlier, the percentage of air at atmospheric pressure in the combustion chamber in relation to the volumetric size of the combustion chamber is called volumetric efficiency.

Next is compression, or squeeze. At the bottom of piston travel during the intake stroke, the intake valve closes. The piston travels past bottom dead center and begins one of its upward strokes. With both the intake and the exhaust valve closed, the upward movement of the piston begins to compress the charge in the cylinder. As the charge is compressed, the temperature of the charge rises. This process is known as adiabatic heating. Adiabatic heating is essential for proper combustion of the air-fuel charge. During the compression stroke, temperatures will rise several hundred degrees.

Power, or boom, is the factor. As the piston approaches the top of the compression stroke, the ignition system fires the spark plug. The firing of the spark plug begins the burn of the air-fuel charge. Momentum carries the piston over top dead center (TDC) as the igniting, heating air-fuel charge begins to expand. As the burning air-fuel charge expands, it drives the piston downward. The power being released by the expanding gases is transferred from the top of the piston to the crankshaft by means of the connecting rod.

Exhaust, or belch, is the fourth factor in engine operation. As the piston swings past bottom dead center (BDC), the exhaust valve opens. The piston starts its upward journey, pushing the remaining burned air-fuel charge out of the cylinder through the exhaust valve. As the piston approaches top dead center the intake valve begins to open, readying the cylinder for the next engine cycle.

It is important to note that on most modern engines, the intake valve opens before the exhaust closes. This is known as valve overlap. Valve overlap accomplishes a couple of things. First, the flow of the exhaust gases out of the combustion chamber creates a low pressure that assists the air-fuel charge in entering the combustion chamber. Second, as the piston starts its trip down on the intake stroke, the exhaust valve is still open. Some of the exhaust gases are drawn back into the combustion chamber. Since the exhaust gases have a very low oxygen and fuel content, the gases that are drawn back into the combustion chamber contribute nothing to the combustion process. However, they do absorb heat during combustion, lowering the combustion temperature. Keeping the temperature in the combustion temperature below 2,500 degrees Fahrenheit reduces dramatically the production of environmentally hazardous oxides of nitrogen.

The cylinder head is as important to the proper performance of an engine as the nose and throat are to a marathon runner. Can you imagine running twenty-six miles with your nose and throat congested? Can you imagine playing an entire rugby championship game with a cotton ball in each nostril and a sock stuffed in your mouth? Asking an engine to perform to its maximum potential when the cylinder heads are not capable of breathing adequately is much the same. All the mail-order, bolt-on parts in the world will not add a single horsepower if the heads do not permit the necessary air into the combustion chamber to supply that power. And all the bolt-on parts in the world will not add a single horsepower if the cylinder head will not allow the burned exhaust gases to be purged properly from the cylinder by the rising piston. When it comes to increasing engine performance, the real key is increasing the amount of combustible air present in the cylinder, and the key to that is increasing the flow capabilities of the cylinder head.

What Does the Cylinder Head Do? 2

Pathway for Intake and Exhaust Gases

The cylinder head forms the lower and upper end of the intake system. The volume of air that flows through both the intake and exhaust system depends on how much air can flow through the point of greatest restriction. If 300cfm is all that can flow through the exhaust system, then 300cfm is all that can flow through the intake. This could be compared to the weakest link in a chain. Therefore, you must ensure that the point of maximum restriction in the cylinder head is at a minimum or removed from the head altogether.

Mixing Venue for Air-Fuel Charge

In both carbureted and fuel-injected engines the fuel and air travels through the intake manifold together. In port fuel-injected engines, it travels only a few inches or even a fraction of an inch. The fuel is injected directly on top of the intake valve, and the mixing of the fuel with the air takes place almost entirely in the cylinder head. Other fuel systems introduce the fuel to the air at a central point in the intake system. Once the air-fuel mixture enters the cylinder head, the head must keep the mixture combined and the fuel atomized into small droplets.

Keeping the droplets atomized is not an easy task. When the fuel is introduced into the air stream headed toward the combustion chamber, it is introduced in relatively large droplets. These droplets must remain in suspension in the air stream until they enter the combustion chamber. If the surface of the intake valve

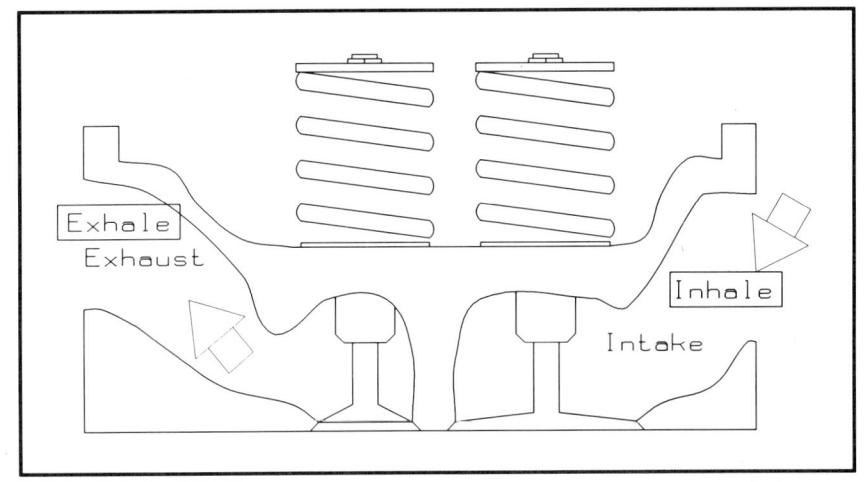

One of the main duties of the cylinder head is to provide a pathway for the air to the combustion chamber and a path for the exhaust gases to exit the combustion chamber.

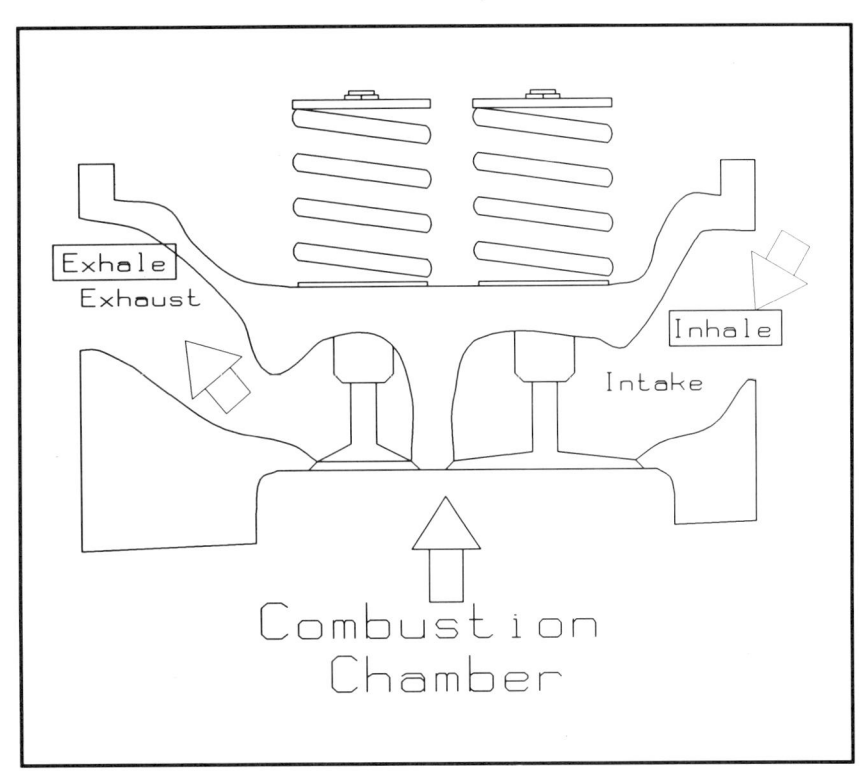

The cylinder head also provides a mixing venue for the air/fuel mixture. In most gasoline engines the largest part of the combustion chamber is in the cylinder head. Contrast this with a diesel engine, in a diesel the largest part of the combustion chamber is in the piston head.

15

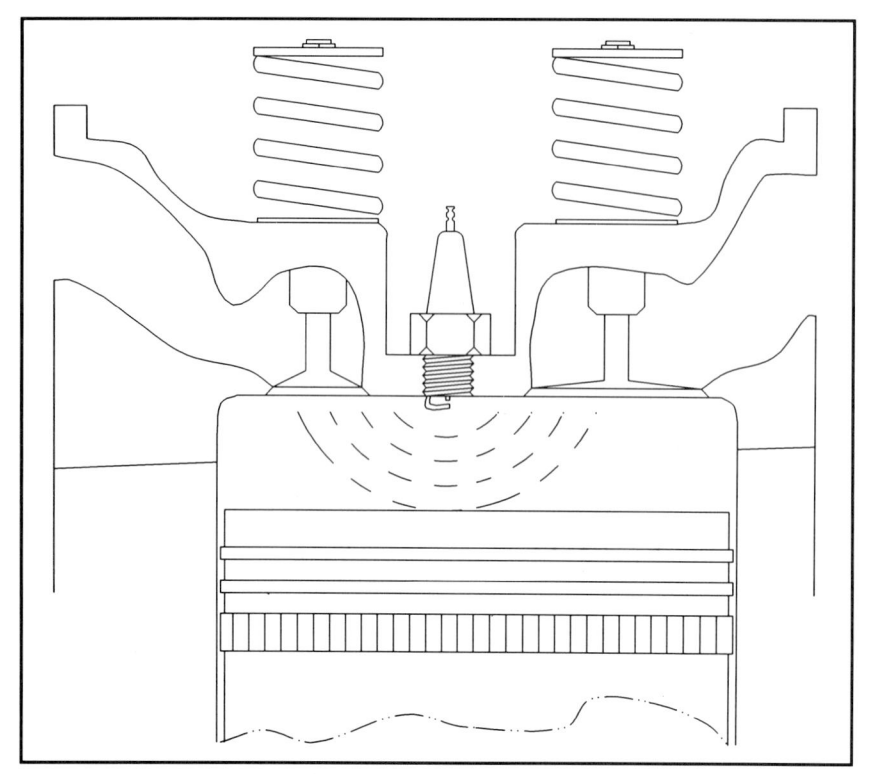

The upper part of the combustion chamber is where the flame front begins. After the spark plug sparks, the flame front then propagates in waves of expanding gases.

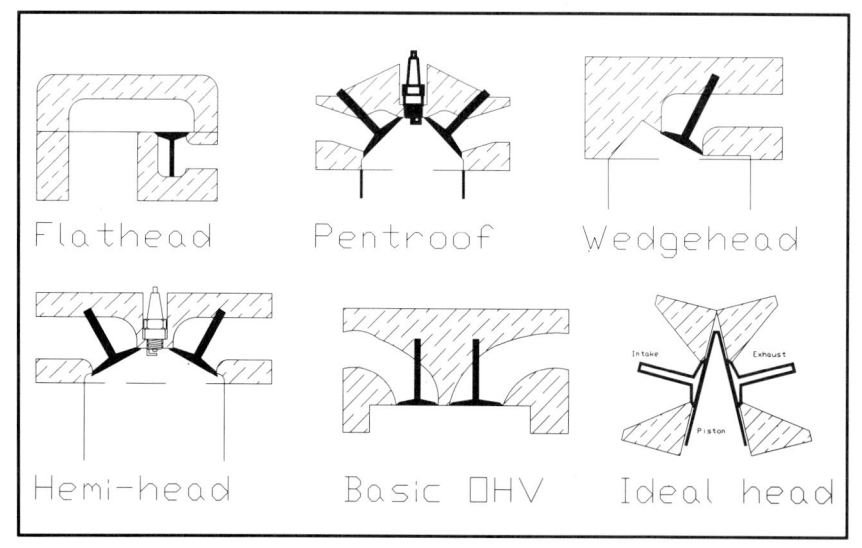

The design of the combustion chamber "shapes" the charge as the mixture explodes and forces the piston down. Above are several combustion chamber designs. In the flathead design the movement of intake and exhaust gases is awkward. In the pentroof and hemi design there is good cross-flow and good shaping of the flame front, attempting to concentrate it in the center of the piston. In the wedgehead design the flame front is fairly well protected but the air flow is awkward. In the "ideal head" air flow potential is excellent, but the concentration of the flame front is very poor. In fact the better the design of the head for air flow, often the poorer the design for flame front concentration.

pocket is too smooth, the air will flow smoothly through the valve pocket. As the air flows smoothly, the heavier fuel droplets will fall out of suspension. One of the most common operations done on modified cylinder heads is to port and polish. Often the polishing creates a smooth surface, which decreases turbulence in the intake port and allows the heavy fuel to fall out of suspension.

Upper Combustion Chamber

The combustion process actually begins in the upper part of the combustion chamber. The shape, size, and configuration of this area contributes to the power that can be created in the cylinder. Much research and development has been done on the subject in the last few decades. From the mid-1950s to the early 1980s, there was little or no improvement in the design of the cylinder head combustion chamber—at least in North America. However, the Asians and Europeans made great strides during the same period. By the early 1980s, many European and Asian cylinder head and combustion chamber designs on engines sold to the public matched the design of those used on racing engines from only a decade or so earlier.

Site for Start of Combustion

The cylinder head is the site for the beginning of the flame front. Gasoline engine cylinder heads have a recessed area that provides clearance between the top of the piston and the bottom of the valves. This area is called the combustion chamber. When the air and fuel have been combined, sealed, and compressed in the combustion chamber, the spark plug fires and a flame front begins to propagate away from the spark plug. As the flame front grows, the nitrogen—the noncombustible element of the air that was drawn into the combustion

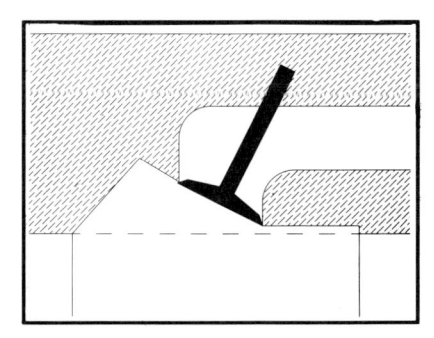

The wedgehead design is the most common combustion design. In this illustration the implication is the intake and exhaust ports are on the same sides of the head. This may or not be true. If the ports are on opposite sides of the head the gas flow through the head still has to perform two 180 degree turns.

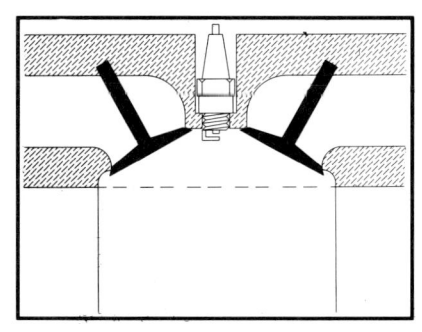

The hemi was the dream engine of the 1950s and '60s. The hemi features good cross-flow. The domed combustion chamber requires the use of a domed piston to keep the compression ratio high. This design has long been a favorite of performance engine builder.

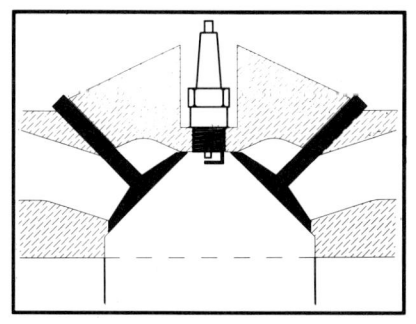

The pentroof design is a lot like the hemi but without the domed characteristic of the hemi. There is very good cross-flow and high compression is maintained through the use of raised-top pistons. This design is very popular in multi-valve engines.

chambers—begins to expand. As the nitrogen expands, the piston is driven downward.

Guide for Combustion Gas Expansion

Since combustion begins in the cylinder head, its shape provides for the initial "shaping" of the explosions in each cylinder. This shape can either provide for smooth gas expansion or contribute to multiple shock waves that interfere with one another. Shock wave interference wastes energy and increases the likelihood of detonation and its resulting power loss.

Combustion Chamber Types

There are four basic combustion chamber types. The first is the wedge. Although there are older combustion chamber designs, the wedge is the oldest of those covered in this book. The wedge head design dates to the 1950s, when manufacturers first began to turn the family sedan into a land rocket. The wedge is the classic overhead valve combustion chamber.

Another combustion chamber design is the hemispheri-cal. They say that anyone who remembers the sixties missed them. That may or may not be true; however, anyone who remembers the sixties and was into performance cars remembers the Chrysler hemis. The unique—for its time—cross-flow design of the cylinder head allowed for phenomenal intake velocity and exhaust gas scavenging. Shortly after graduating from high school, one of our car-head gang "lowered himself" to buy a Toyota Corolla with a 1600cc engine. Randy was immediately blacklisted, left out of the gang's activities. Then we learned that the

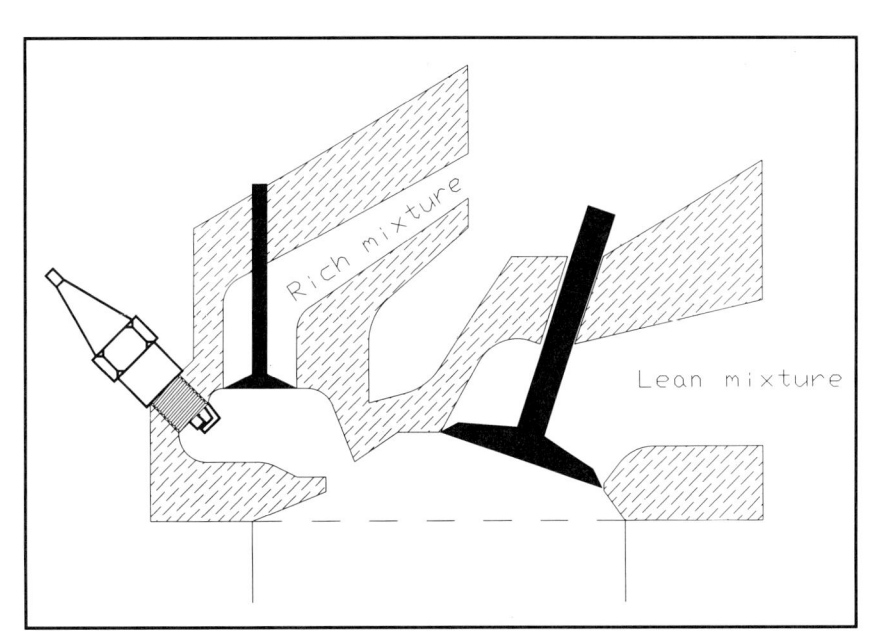

The stratified charge cylinder head used by Honda, Mitsubishi, and others features two intake valves. The larger intake valve allows a very lean (or stratified) charge to enter the combustion chamber. The smaller intake valve admits a very rich mixture into a pre combustion chamber. This rich mixture is ignited by the spark plug. The rich mixture burns slowly and keeps the stratified mixture in the main combustion chamber burning. This technology means lower emissions and improved fuel economy.

17

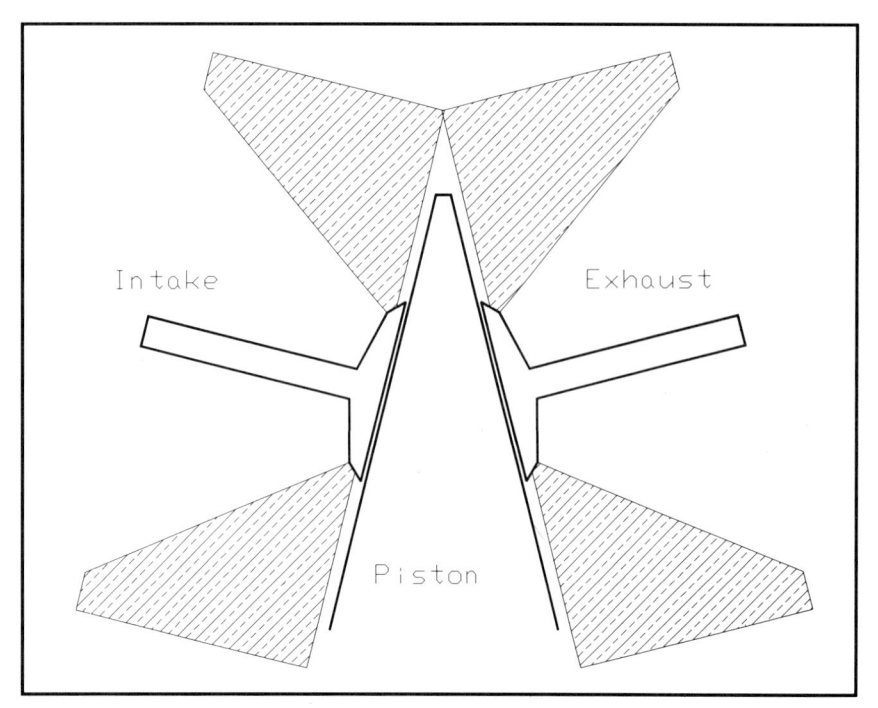

This was called the "ideal" head in an earlier illustration. It is ideal only with respect to flow of intake and exhaust gases. Upon ignition the flame front would be directed down the sides of the pistons. Good flow but lousy shaping of the flame front.

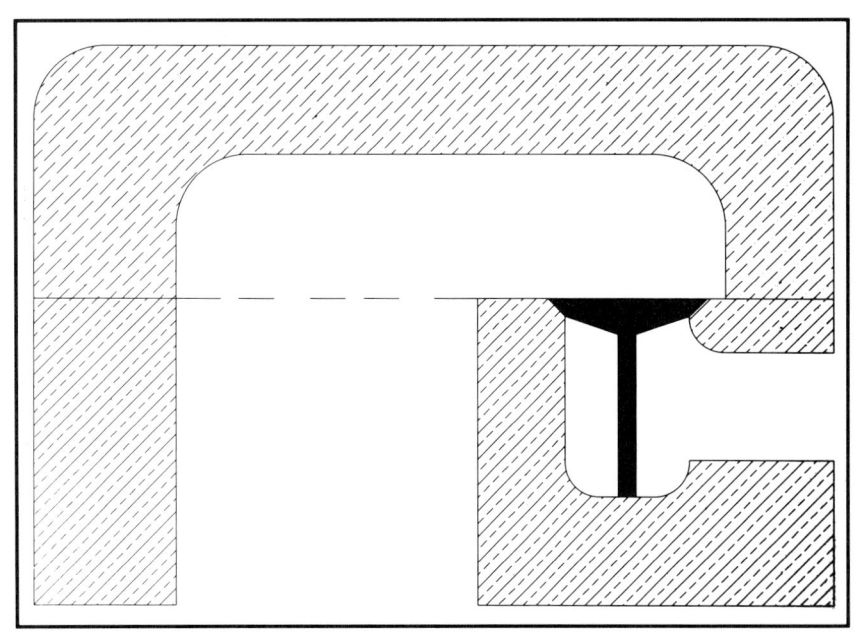

The flathead design dates to the earliest days of the automotive industry. Air entering the combustion must do a 180 degree turn. Exhaust gases exiting the combustion chamber must also exit by means of a 180 degree turn. Additionally the piston never fills the combustion chamber, and the result is inherently low compression ratios.

1600cc Toyota engine had hemispherical combustion chambers. Randy was immediately repatriated. He was the only guy in the group to have a hemi! Since that time, the hemispherical combustion chamber has been used to increase both engine power and fuel economy.

A third combustion chamber design is called the stratified charge. The best known of these types is the Honda CVCC (Centrifugal Vortex Combustion Chamber). The spark plug ignites a rich charge in a pre-combustion chamber. This mixture burns slowly and provides a heat source for the continued ignition of a lean (stratified) mixture in the main combustion chamber. The biggest advantage of this design is very low emissions.

One of the latest combustion chamber designs, the pent roof, is used on many multivalve engines. The design provides for excellent cross flow of intake and exhaust gases.

Tools

3

Doing your own cylinder head work will require that you acquire a variety of tools. Whether you purchase, rent, or borrow these tools will likely depend on your budget.

Compression Tester

The compression tester is a pressure gauge with a check valve. When installed in a spark plug hole, the compression gauge is able to record and display the amount of pressure created in the cylinder as the engine is cranked.

To get ready to run a compression test, remove all of the spark plugs, disconnect the primary power supply to the ignition coil, and block the throttle at wide open. Install the compression gauge in the number one spark plug hole. Then crank the engine twelve revolutions. Listen to and count the "puffing" noises. Record the compression reading, and move to the number two cylinder. Repeat the process on each of the cylinders. When done, compare the compression readings. If any of the cylinders are considerably different than the rest, repeat the test. For the second test, squirt a teaspoon of oil into each cylinder before testing it. This is called a wet compression test.

Cylinder Leakage Tester

The cylinder leakage test is superior to the compression test, although its goals are the same. It also requires more equipment. With the piston at top dead center (TDC) of the compression stroke, connect the cylinder leakage tester to cylinder number one. Measure the percentage of leakage. Also use a piece of heater hose held to the ear to determine where the leakage is. Stick the heater hose in the open bore of the fuel injection throttle assembly or carburetor. If there is a great deal of air escaping through there, you have an intake valve problem. Stick the hose in the exhaust. If there is a good amount of air escaping, then you have a bad exhaust valve. Observe the radiator coolant. If there are a lot of bubbles in the coolant, you have a blown head gasket, or a cracked head or block. Next remove the oil filler cap. Place the heater hose in the oil filler neck or valve cover opening. If you hear an excessive amount of air escaping, then the rings are allowing the air to bypass. Now place the hose over the spark plug holes of the adjacent cylinders. Again, if you hear a significant amount of air escaping from one of the adjacent cylinders, then the head gasket is blown between where the cylinder is being tested and where the air is escaping.

Be aware, however, that in *any* case there will be a certain amount of air leakage in these areas. But when some cylinder damage is present, leakage will be more significant. The difficult part of the cylinder leakage test is determining how much leakage is *too much*.

Opinions differ on this topic. Manufacturers such as Honda say that leakage in excess of 5 percent is too much. Some professional technicians feel than 15 percent is acceptable. The truth of what is acceptable lies somewhere in between. Being realistic, you have gone to the trouble of performing the cylinder leakage test because there is an obvious problem. Generally speaking, the types of problems that would be caused by 5 to 10 percent leakage

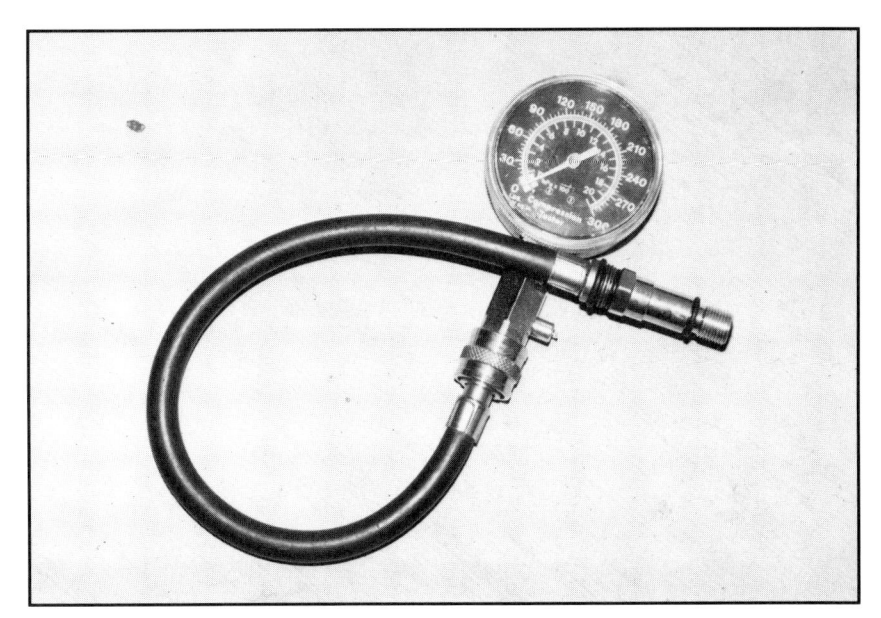

The compression gauge has long been a primary diagnostic tool for checking how well the piston, the rings, and the valves do their job.

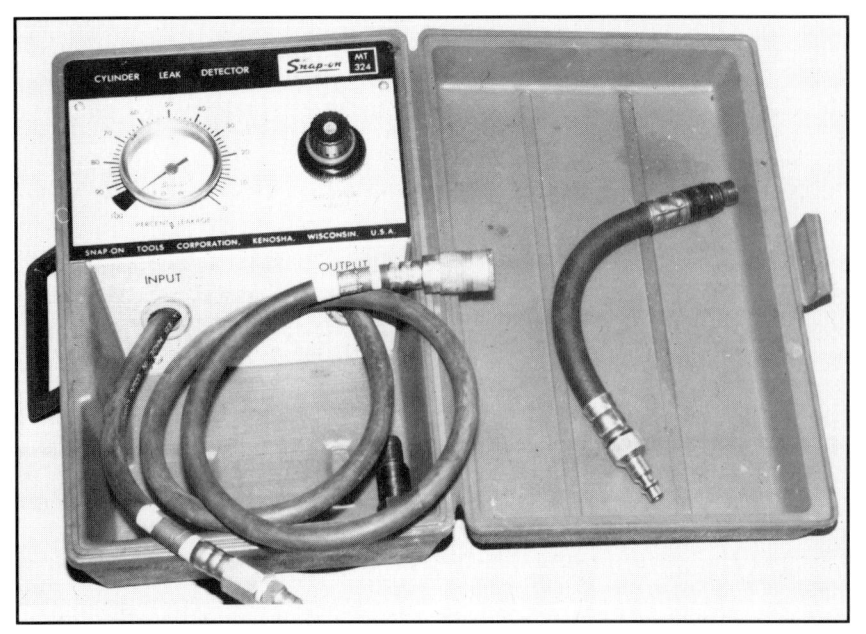

The cylinder leakage tester is considered by many technicians to be superior to the compression tester. This tool forces compressed air into the cylinder. The gauge then measures the amount of air leaking out of the cylinder. As the air leaks out, the hissing sound of that leak can be sought out by the technician. If the hissing is heard in the radiator, there is a blown head gasket. If the hissing is heard in the exhaust, the exhaust valve is leaking. If the hissing is heard in the throttle bore, the intake valve is leaking. If the hissing is heard in the oil filler neck, the rings are leaking or the piston has a hole in it.

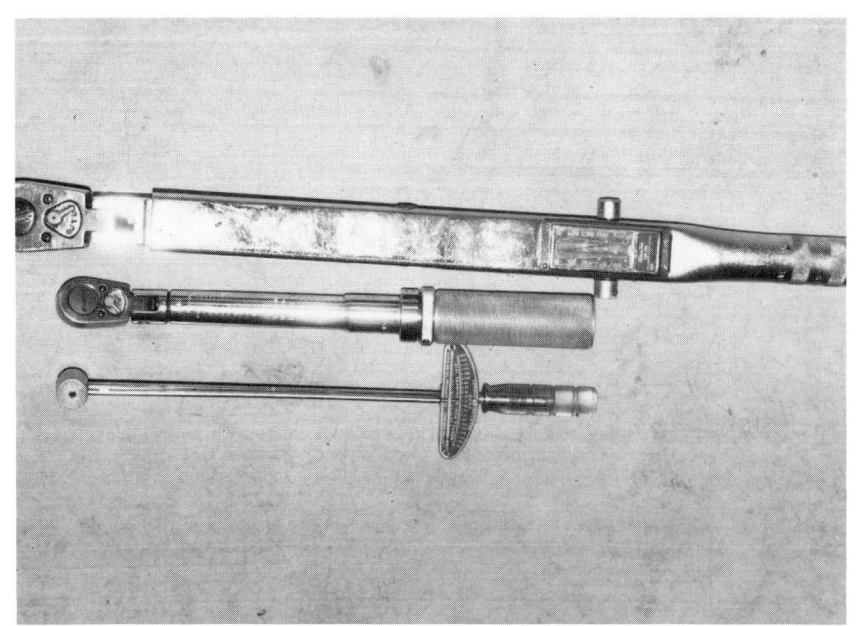

The torque wrench is the most important tool in doing any engine work. Shown are two click type torque wrenches and one beam torque wrench. While the author rebuilt many engines and heads using a beam torque wrench, there is no doubt that the click type is far superior.

would not be noticed. You are looking for a problem resulting from cylinder leakage that is probably greater than 15 percent.

Oil Pressure Gauge

The oil pressure gauge is a mechanical gauge that screws into one of the oil galleries in the engine block. When the engine is started, the oil pressure should rise to the proper specification or higher.

Torque Wrench

The most essential tool for rebuilding engines is the torque wrench. All bolts and other fasteners stretch when they are tightened. If they are tightened too much, their threads will strip or they will shear. If they are not tightened enough, the bolts will loosen. So, spare no expense when you purchase a torque wrench.

There are two types of torque wrenches available, the beam type and the click type. The beam type uses a pointer that indicates the amount of torque on a scale attached to a flexible bar. As torque is applied to the bolt, the flexible bar bends. The amount of bending in the bar indicates the amount of torque being applied to the bolt. The beam torque wrench was the industry standard for decades, and in highly skilled hands is still an adequate tool. Its biggest weakness is that it is not consistently accurate, even in the most skilled hands.

The click torque wrench is far more accurate, especially for the occasional user. The desired torque is dialed in. When the prescribed torque is reached, a click is heard and felt, indicating to the user to stop tightening the bolt. These torque wrenches feature a high degree of accuracy. Although the click torque wrench can cost significantly more than its beam counterpart, it can be money well spent. One incorrectly torqued bolt can result in having to completely redo an engine rebuild.

This is both expensive and time consuming.

Valve Guide Drift

Some cylinder heads have valve guides that are pressed into place. Although it is advisable to have these guides replaced by a qualified machinist during a valve grind, you can replace them yourself with a special set of drifts. These drifts not only have a surface for pressing the guide into place, but they also have an undercut area that fits into the guide to prevent it from collapsing on itself. Anytime the valve guides are replaced, a valve grind should be performed. For this reason, unless you plan on doing the valve grind yourself, let the machinist replace the guides.

Micrometer

An automobile engine is made up of dozens of closely machined parts. Measurement of these parts requires precision to the thousandths of an inch. The micrometer provides this level of accuracy. A basic set of micrometers consists of a 0-1in, 1-2in, 2-3in, and 3-4in. When working with larger engines, such as big-block and truck engines, they may require a 4-5in micrometer or larger. A micrometer is a precision instrument that should be handled with extreme care.

In spite of its high level of accuracy, the micrometer is a fairly simple device. The basic parts include the spindle, the thimble, the anvil, and the barrel. The spindle is threaded at a standard of forty threads per inch. Rotating the thimble moves the spindle 0.025in. On the standard micrometer there are twenty-five graduations around the thimble. This configuration allows an accuracy of 0.001in. The barrel of the micrometer is marked off in 0.025in units. There are numerical markings every four barrel marks. Each marking, therefore, indicates a movement of 0.1in.

Many cylinder heads have replaceable valve guides. These guides are pressed into the heads. Removal and installation will require drifts/drivers such as these.

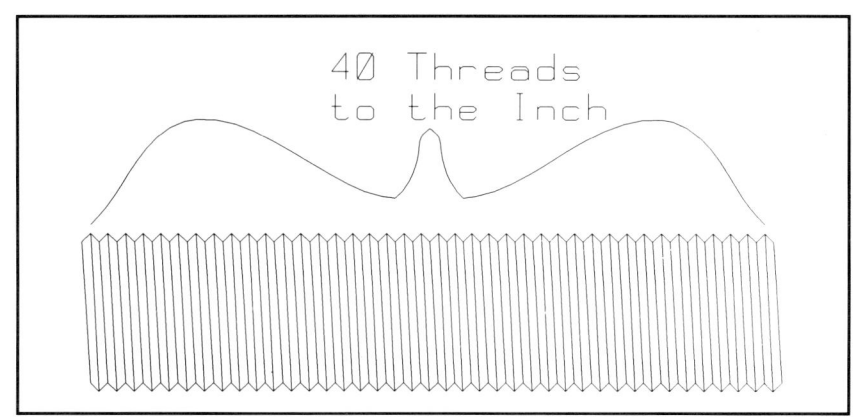

The micrometer works on a very simple principle. The spindle is threaded with 40 threads per inch. Every rotation of the barrel on the micrometer equals a movement of 0.025in.

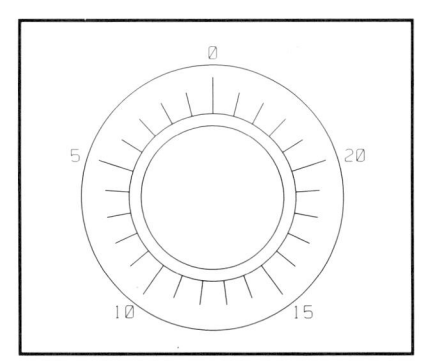

One rotation of the barrel equals 1/40in. One twenty-fifth of a rotation equals 1/1,000in of movement.

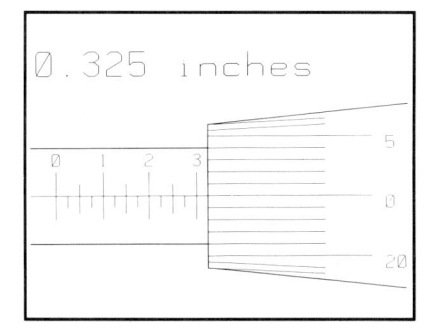

This close-up of a micrometer shows a measurement of 0.3in plus one additional rotation of the barrel for a total of 0.325in.

21

Micrometers are precision measuring instruments. Capable of measuring to an accuracy of at least 1/1000in, they can accurately measure the thickness of a human hair.

Less accurate than the micrometer, a vernier caliper is often faster and easier to use. The plastic model works well for measuring valve adjusting shims on overhead cam engines.

Using a micrometer involves placing its anvil and spindle against opposite ends of the object to be measured. Rotate the thimble until the spindle makes firm, but not heavy contact with the object. Read the highest numerical marking visible on the barrel. This will indicate the nearest 0.1in. Counting the barrel markings beyond the visible numerical mark will indicate the nearest 0.025in. Observe the twenty-five marks on the thimble and how they align with the line on the barrel. This will indicate the measurement to the nearest 0.001in.

A quality micrometer should be treated as the precision instrument it is. Overtightening when taking measurements can stretch the frame, decreasing the accuracy of the micrometer. Often this can be remedied by adjustment using a gauge block. Dropping the micrometer can damage it beyond adjustment. Top technicians may be more inclined to loan you their spouses than their micrometers.

Inside Micrometer

The inside micrometer works just like the outside micrometer already described; however, its design allows it to be used to measure the inside of a bore.

Telescoping and Split-Ball Gauges

When the cylinder or bore is smaller than 2in, the inside micrometer cannot be used. Extremely small bores require the use of split-ball gauges. When inserted into the bore of a component such as a valve guide, the split-ball can be adjusted to the guide diameter. When removed from the bore, a standard micrometer can be used to measure the size of the split. This will be the guide diameter.

Telescoping gauges are available in sizes up to several inches. They consist of spring-loaded rods that make contact with the sides of the cylinder or bearing bore to be measured. When removed, the length of the telescoping rods can be measured with a standard micrometer to determine accurately the bore diameter.

Note: Using micrometers, inside micrometers, and split-ball and telescoping gauges requires a little skill. The measuring instrument should be extended until it makes firm, but gentle contact with surfaces being measured. Gently move the measuring device back and forth to ensure firm contact.

Dial Indicator

The dial indicator consists of a precision gauge, usually marked in 0.001in, which is moved by a plunger. When clamped or mounted firmly to the block or cylinder head, the indicator can be used to measure the amount of front-to-rear movement or end play in the camshaft or crankshaft. Not limited to this job, the dial indicator also can be used to measure minute movement in virtually any component.

Additional jobs for the dial indicator include checking the flatness and run-out of rotating objects like the flywheel.

Feeler Gauge

Feeler gauges are thin strips of steel or brass machined to very close tolerances. They are used to measure the gaps between components. Piston ring side clearance, crankshaft thrust clearance, and valve lash are among the gaps that can be measured with feeler gauges.

Plastigauge

Plastigauge is made of thin strips of plastic and is used to measure the rod and main bearing clearances. Remove the bearing cap and tear off a piece of Plastigauge approximately the length of the bearing width. Place the strip of Plastigauge on the bearing journal, then install and torque the bearing cap. Remove the bearing cap and compare the compressed width of the Plastigauge to the scale on the package. The compressed width indicates the oil clearance between the journal and the bearing.

Tachometer

The tachometer is used to measure engine rpm. While this tool is not essential for the actual engine rebuilding process, it is essential for adjusting the curb idle speed during the final tuning process.

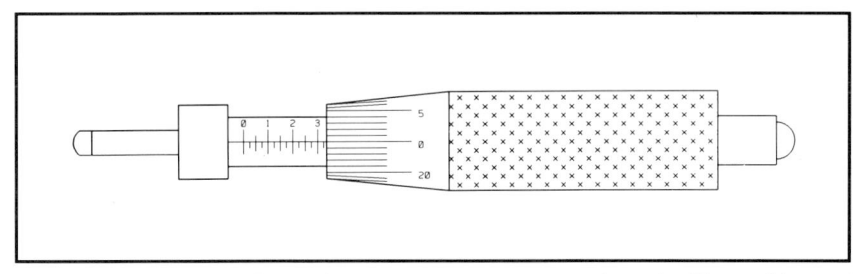

The inside micrometer is used to measure the internal size or diameter of components. It works like and is read like an outside micrometer.

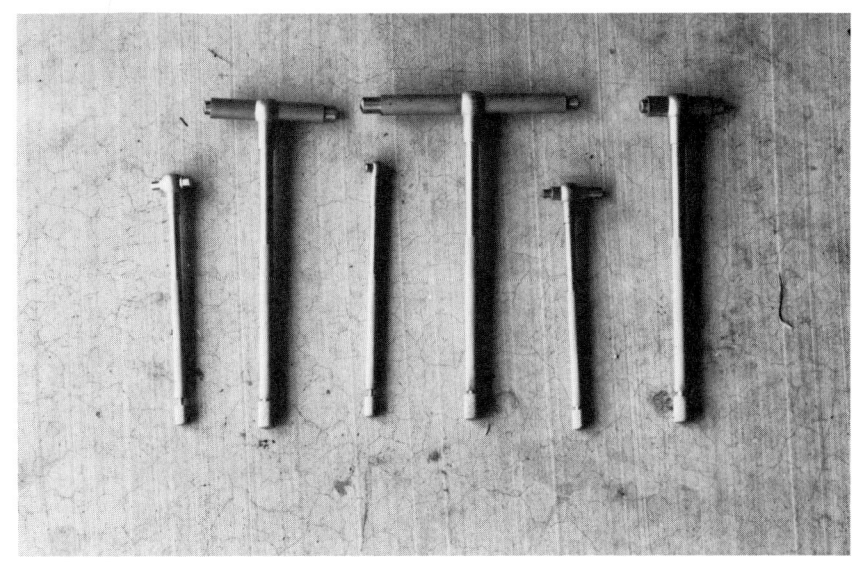

A more versatile and less expensive tool is the telescoping gauge. The inside diameter is measured with the telescoping and then the telescoping gauge is measured with a micrometer.

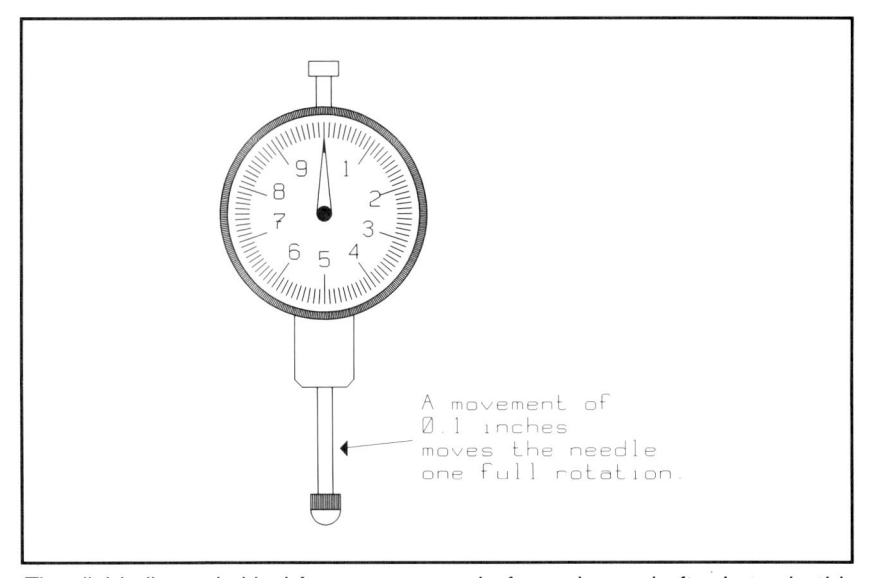

The dial indicator is ideal for measuring small movements. Often they are used to measure end play in crankshafts and camshafts. Later in this book it will be used to locate TDC and accurately set the cam timing.

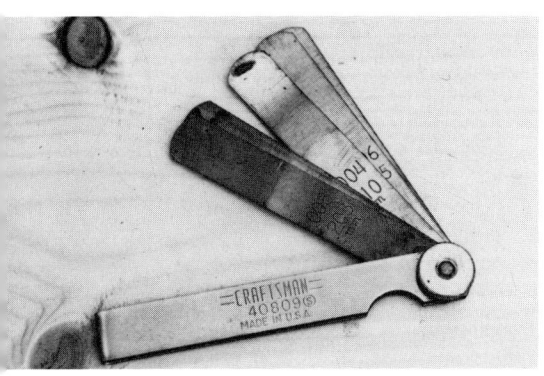

Feeler gauges are necessary to adjust the valves on most applications. They are used to measure the clearance between two surfaces.

A rather simple tool and relatively inexpensive, the coolant system pressure tester checks the cooling system for leaks. One of the more common head problems is blown head gaskets. This tester can reproduce the pressures inside the cooling system of a hot engine. The technician can the check the cooling system and engine for leaks.

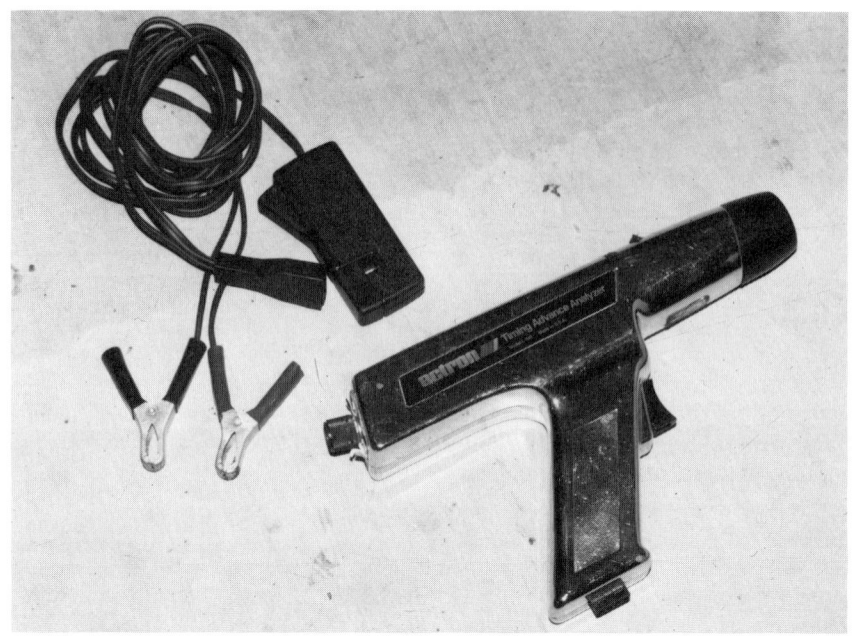

In my career as an automotive technician I have owned more than a half dozen timing lights. A little known fact is the affinity of the cables for the radiator fan. This timing light has three of my favorite features: relatively inexpensive, replaceable cables, and a knob to adjust the flash to check timing advance.

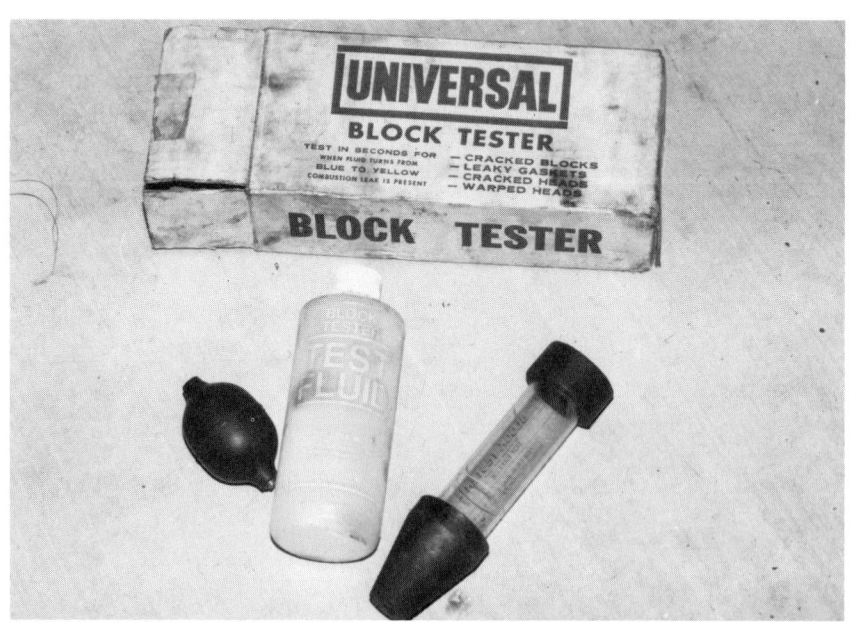

When it comes to finding blown head gaskets, which of course is the best way to justify major head work to your spouse, the block tester is the most definitive. The kit comes with a bottle of blue test fluid. This bottle was emptied trying to justify performance work to my wife on the heads of her Cadillac. The engine is started and brought to operating temperature. The plastic tube to the right is filled about 1/4 full with the blue fluid then placed in the neck of the radiator and the rubber bulb to the left is used to pump or draw radiator vapors into the tube. If these vapors contain combustion gases (hydrocarbons), then the fluid will change from a blue color to a greenish-yellow color.

Timing Light

Like the tachometer, the timing light is a tune-up instrument, not an engine rebuild necessity. The timing light is used to synchronize the primary ignition system to the position of the crankshaft.

Radiator Pressure Tester

The primary task of the radiator pressure tester is to locate leaks in the cooling system. When installed on the radiator neck, the pressure tester pressurizes the cooling system and lets you look for where coolant will seep or leak out. An added benefit is the tester's ability to locate a leaking external head gasket and leaking freeze plugs.

Chemical Block Tester

Filled with a blue fluid and inserted into the filler neck of the radiator, the chemical block extracts some of the vapor in the radiator. The presence of the hydrocarbons in the radiator vapors indicates leakage from the combustion chamber into the radiator. This leakage could be a result of a blown head gasket, or a cracked cylinder or cylinder head. If hydrocarbons are present in the vapors, the blue fluid will turn yellowish green.

Valve Spring Compressor

This tool looks like a large C-clamp with a set of hooks on one end to cradle the valve spring. When tightened, the valve spring is compressed, which allows the valve keepers to be removed. Although not absolutely necessary during the disassembly process, the valve spring compressor is indispensable during the reassembly process.

Some overhead cam engines have the valves recessed in the head. These engines require special valve spring compressors.

Ridge Reamer

The ridge reamer is used to remove the ridge that builds at

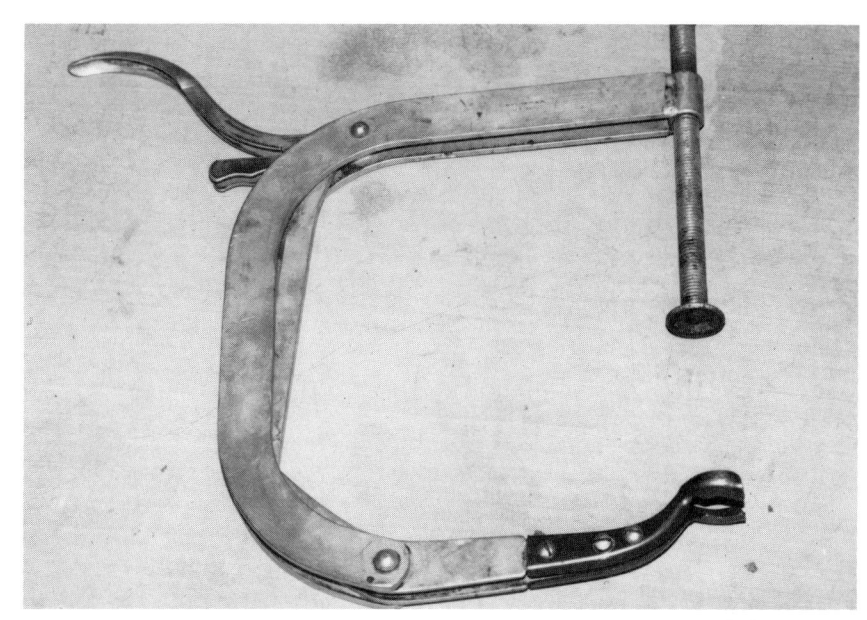

One of the necessary tools for doing head work is the valve spring compressor. This one I ordered from the Montgomery Ward catalog in 1972. After more than 20 years of service it still serves well. A lot of money can be spent on air-operated valve spring compressors, yet they really have limited advantage over this type.

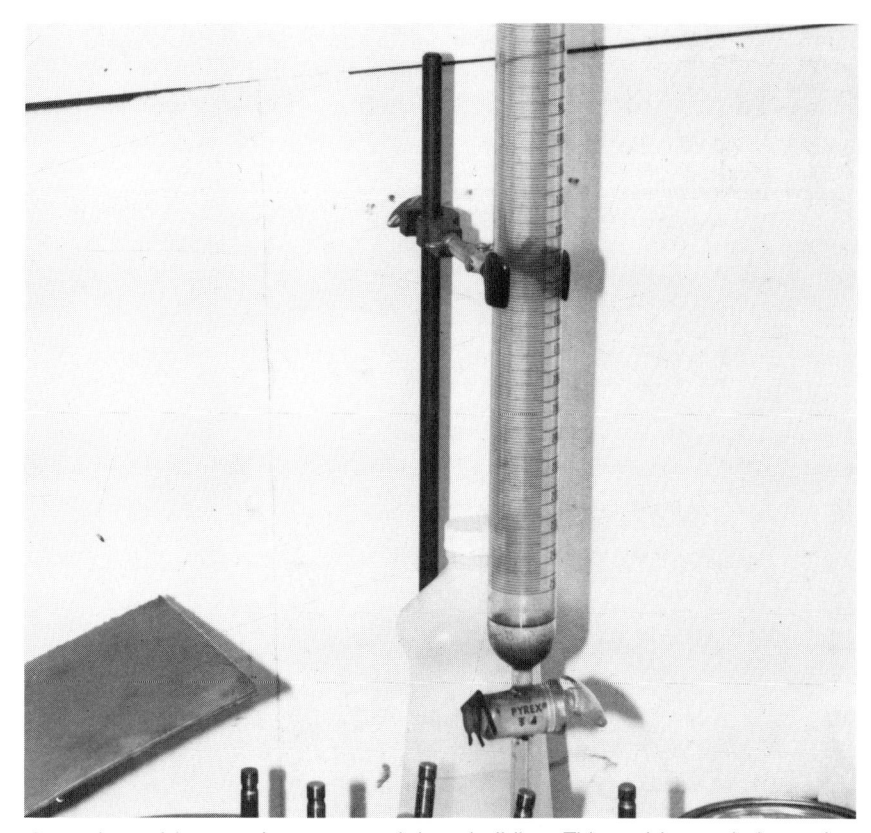

A graduated burette is an essential tool to the performance engine rebuilder, yet you will never find this tool in a shop that does regular engine rebuilding. This tool is needed to calculate actual compression ratio and to verify that the cylinder head combustion chambers are all equal in size.

25

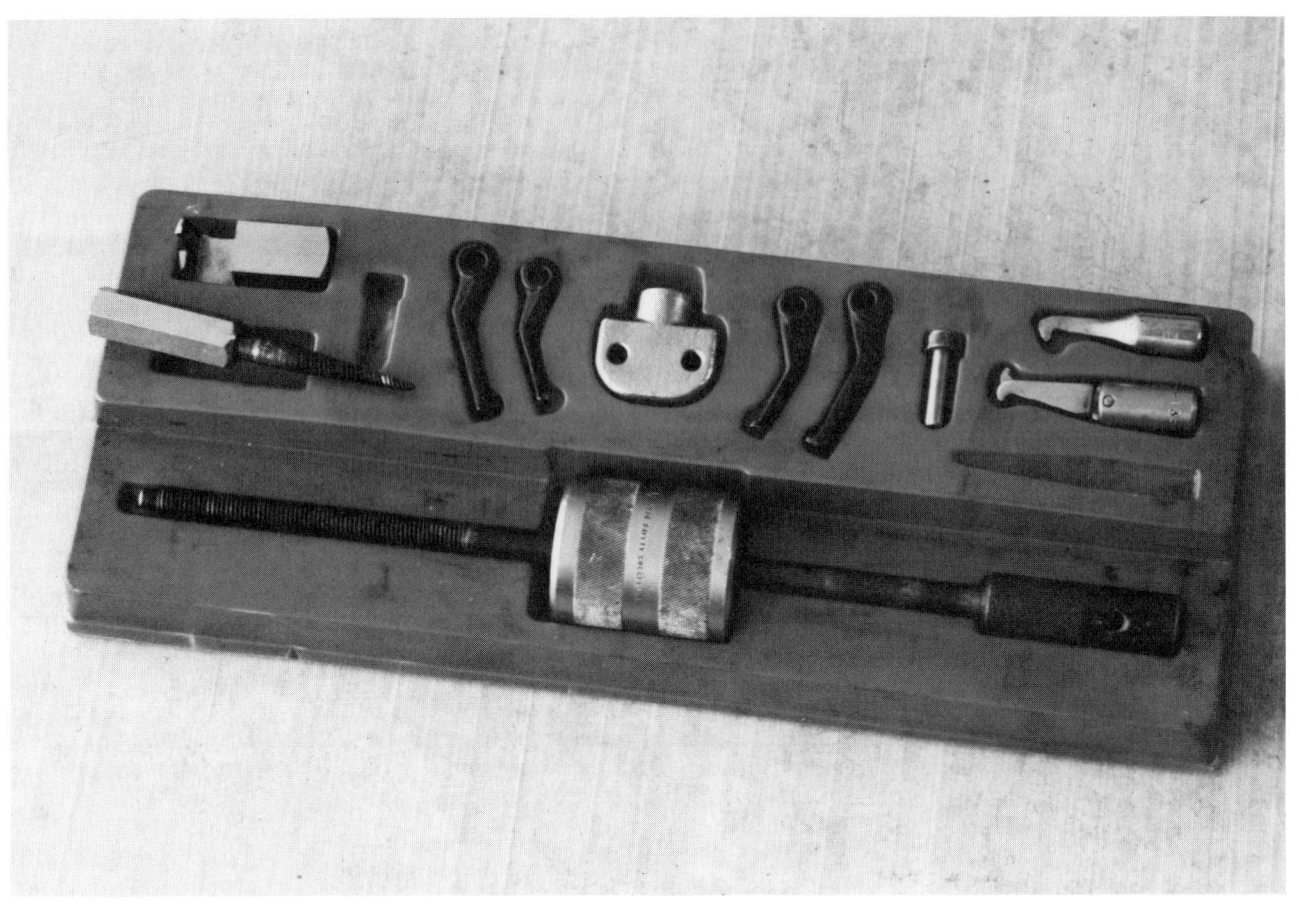

This is a tool that no one ever needs until they need it. A slide hammer is used to remove everything from freeze plug to valves that have been jammed in the guide by a catastrophic failure.

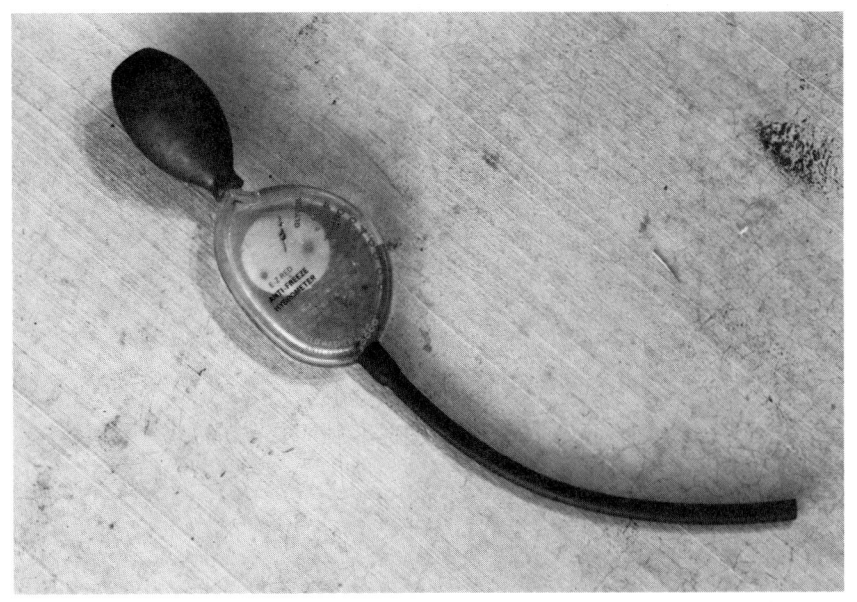

Many people still think of coolant as merely antifreeze. In reality today's engine require year-round coolant. Modern coolant helps transfer heat from the head and other engine parts to the radiator and eventually to the air.

the top of the cylinder walls. This ridge forms during thousands of miles of operation as the rings move tiny bits of metal up the cylinder walls, forming deposits at the top of the cylinder.

Valve Adjusting Clips

These are spring steel clips that fit over the end of the rocker and pushrod on overhead valve (ohv) engines. Ohv engines equipped with hydraulic lifters must have the valves adjusted with the engine running. If these clips are not used, adjusting the valves with the engine running—a necessary procedure on many engines—can be very messy.

CC Gauge

The CC gauge is known to chemists as a graduated burette.

A burette is a glass tube marked in milliliters. At the bottom of the tube there is a petcock. When the burette is filled with mineral oil and slowly drained through the petcock into the combustion chambers of the cylinder head, it is possible to measure the volume of the chambers. When blueprinting an engine, the equality of combustion chamber volume is critical to achieving equal power from each of the cylinders.

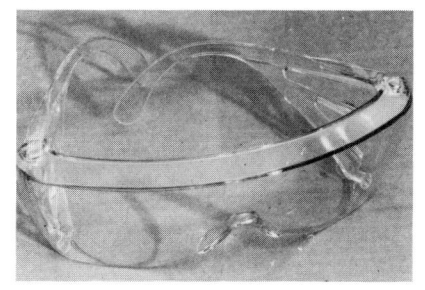

Required Hand Tools

Make sure that you have a set of combination wrenches ranging in size from 1/4in to 1in. A 3/8in drive SAE (Society of Automotive Engineers) and 1/2in drive SAE socket, ratchet, extension set, and miscellaneous screwdrivers, pry bars, and hammers. If the engine is in a foreign car, you will need metric wrenches and sockets. If the car is a late-model domestic, you will need both SAE and metric tools.

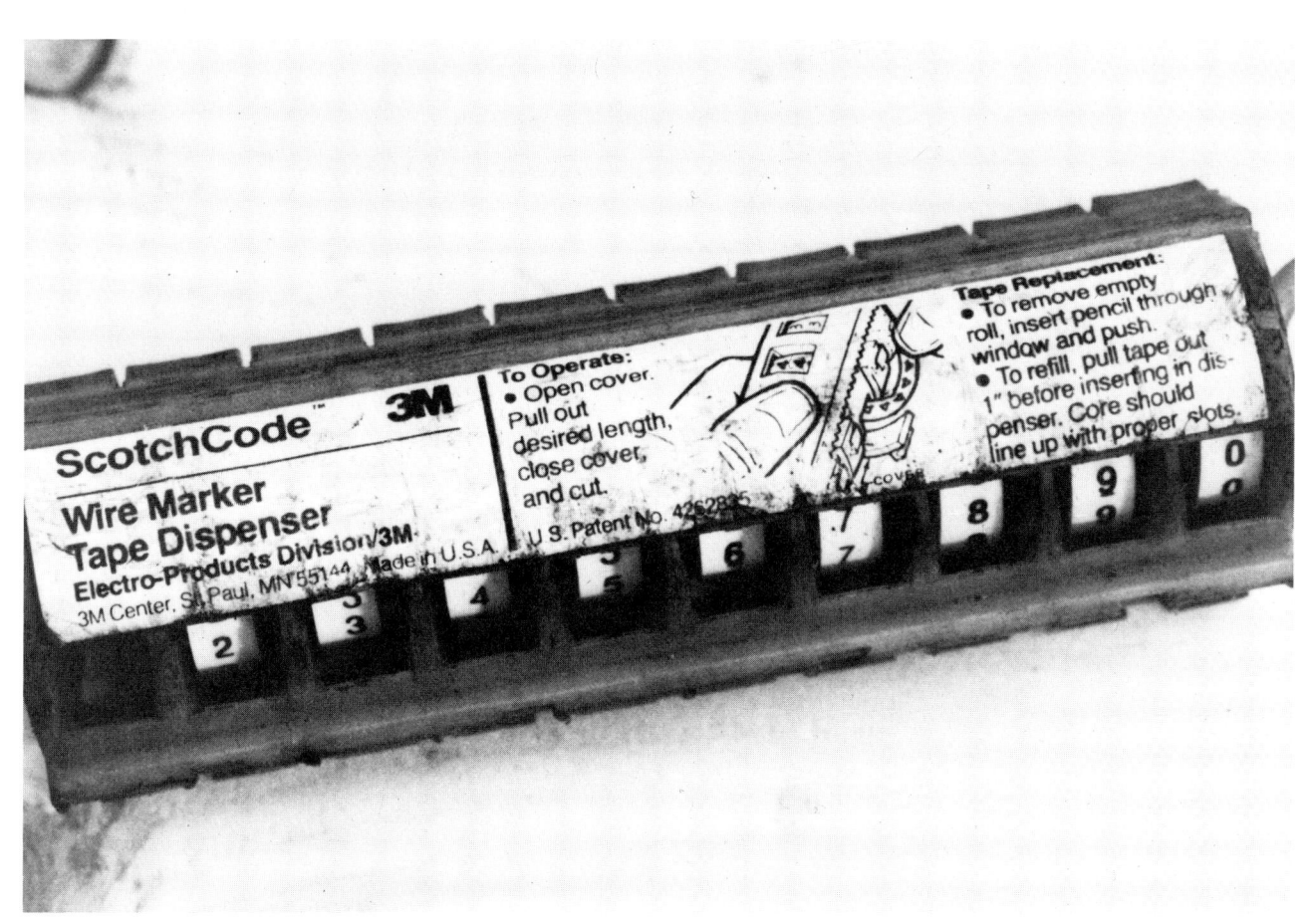

This is possibly the most important tool you can use when doing any engine work. Safety glasses may seem uncool, but avoidable blindness is definitely uncool.

This is a set of commercially available head stands. The set I have was constructed by welding two punches on a couple of pieces of bar stock. The tapered shafts fit into the head bolt holes of the head. Using stands prevents the machined surface of the head from being laid on the bench. This reduces the possibility of damage to this surface.

These numbering tapes are handy to prevent crossing plug wires, vacuum lines, and wiring. Place numbers on the wires and on the connection points. This eliminates the need to think when it comes time to install the rebuilt heads.

Fasteners

4

Bolts

Bolts are measured by the maximum diameter of the threads and the distance from the bottom side of the bolt head to the end of the bolt. The number of threads per inch (or millimeter) determines the pitch of the threads. On an American engine there are typically two thread pitches, course and fine. The finer the pitch the more threads there are per inch, and therefore, the greater the holding power of the bolt.

There are four sizing standards used, UST, ISO metric, pipe, and Whitworth. Within the UST standard there are UNC and UNF. These standards have been used for decades on American-built cars. The UNC standard refers to course-thread bolts while UNF refers to fine-thread bolts.

Bolts designated with the letter "M" followed by the bolt diameter would be course-thread types. For example, a bolt designated M8 would be a course-thread 8mm bolt. Bolts designated with the letter "M" followed by the bolt diameter, then a multiplication symbol (x) followed by a number would be a fine-thread bolt. For example, M10x1.25 would indicate a 10mm bolt with a pitch of 1.25 threads per millimeter.

Pipe thread is indicated by the letter "G" preceding the pipe size. There is only one pitch available for each diameter.

You are likely to find Whitworth-standard bolts on older English-built cars. These British pipe thread measurements are designated by the letter "R" followed by the pipe size for externally threaded parts such as bolts. "Rp" indicates internal threading. For example, R1/4 is a 1/4in external pipe thread, while Rp1 is a 1in internal pipe thread.

What are bolts made out of? Bolts are made from cast iron, steel, malleable iron, aluminum-magnesium, and a variety of exotic metals. Steel bolts can be electroplated with zinc or cadmium. These platings help to reduce corrosion.

Torquing Bolts

There is an art to successfully torquing bolts. Unfortunately, Murphy's Law seems to come into play here. Rule number one is, *Make sure you tighten the bolt until the head breaks off, then back it off a quarter turn.*

On a more serious note, the primary fasteners used in engine assembly are nuts and bolts. Yet, anyone undertaking major vehicular powerplant surgery needs to be aware that a bolt is *not* just a bolt. There are several grades and types of bolts based on strength and corrosion resistance.

There are three major parts to a bolt: the head (the part where the wrench goes), the shoulder, and the threads.

Based on strength, there are two bolt measuring standards, the SAE (Society of Automotive Engineers) and the ISO (International Standardization Organization). Each organization has specifications and standards for tensile strength, yield point, ultimate strength, and proof psi (pounds per square inch) load.

To describe tensile strength, imagine holding the bolt by its head and attaching weights to the end of it. The tensile strength is the projected weight at which the bolt would break.

The yield point refers to the stress point at which the bolt can no longer return to its original shape. A good comparison would be a pair of men's briefs. As long as the wearer of these briefs remains svelte, the elastic retains

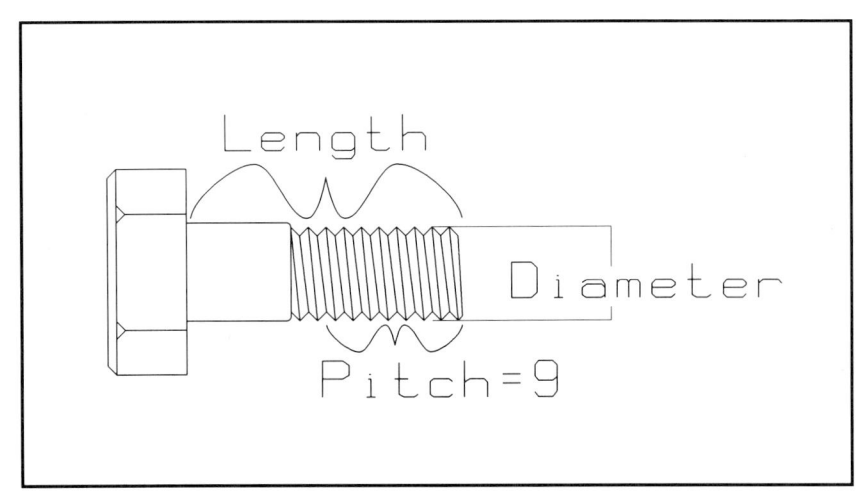

The length of the bolt is measured from the bottom of the head to the end of the bolt. The diameter is measured across the widest part of the threads.

The pitch is the number of threads that occur in a unit of length measurement such as inches or millimeters.

its ability to return to approximately its original size. If, however, several months of waistline increases are followed by a sudden waistline reduction, it is likely the elastic will retain the more corpulent shape.

Ultimate load is the approximate point at which the bolt snaps. As a rule of thumb, this tends to be about 10 percent more stress than the yield point.

Proof psi load is the stress point that is considered to be its typical long-term load value. I used to own a 1965 Volkswagen Beetle with a 1200cc engine. Downhill, with a tailwind, it would achieve 90mph. Little, if any, injury was done to the engine during this short span of maximum stress. However, if I had driven the car flat out for several hours, the engine would have received extensive damage. By driving the VW at the safe and sane speed of 55mph for its entire life (yeah, right!), there was less stress put on the engine and therefore it lasted much longer. Bolts are typically tightened to about 90 percent of their proof psi load. Torque specifications are in foot-pounds, however, not in psi.

These are the types of SAE-standard bolts, classified by strength:

SAE Grade 2

Often referred to as hardware bolts, these fasteners have the lowest strength qualities of any bolts found in use on the automotive engine. Their tensile strength ranges from 60,000 to 74,000psi of cross section. The proof load of these bolts is between 33,000 and 55,000psi. A typical torque spec for a half-inch, Grade 2 bolt is 59lb-ft. These bolts are recognized by a lack of radial lines on the top of the head.

SAE Grade 5

Grade 5 bolts are found where greater torque values and

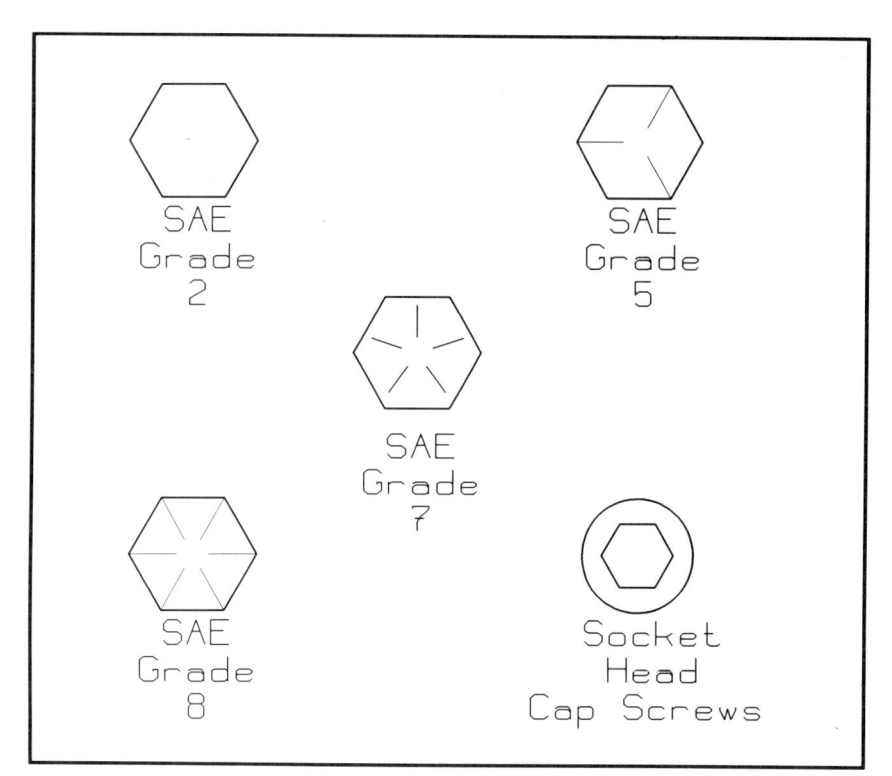

The top of the head of the bolt has markings to identify the relative strength of the bolt. In general the more marks there are the stronger the bolt. However, the socket head cap screw is usually the strongest.

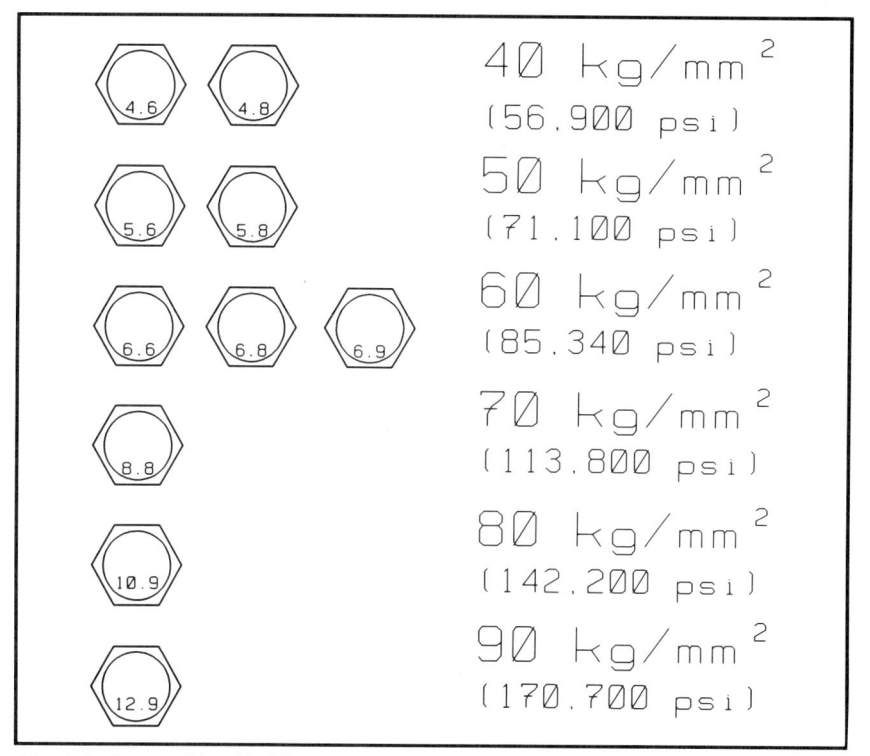

Metric bolts are identified by two numbers separated by a dot. The number to the left of the dot indicates the size of the bolt. The number to the right of the dot represents the strength of the bolt.

an increased need to maintain a torque are required. Their tensile strength at 105,000 to 120,000psi is nearly twice that of the Grade 2 bolt. Their proof load is 74,000 to 85,000psi. A typical torque spec for a half-inch, Grade 5 bolt is 90lb-ft. These bolts will have three radial lines around the head.

SAE Grade 7

The Grade 7 bolt is commonly used in automotive engines. While it has a relatively high tensile strength, its expense is considerably less than the Grade 8. The Grade 7 bolt can be recognized by five radial lines around the top of the head.

SAE Grade 8

When great strength is required, Grade 8 bolts are used. The tensile strength of the Grade 8 is 150,000psi, and the proof load is 120,000psi. A typical torque spec for a half-inch, Grade 8 bolt is 128lb-ft.

Socket Head Cap Screws

When maximum strength is required, socket head cap screws are available. These are designed with a tensile strength of 160,000psi and a proof load of 136,000. A typical torque spec for a half-inch socket head screw is 145lb-ft. These bolts have six radial lines around the head.

That brings us to rule number two: *The more expensive the component into which you are tightening the bolt, the greater the probability of stripping the threads.*

The International Standards Organization (ISO) rates metric bolts. ISO ratings include tensile strength and proof load. The tensile strength is measured in (kg/mm^2).

If the tensile strength of the bolt is $40kg/mm^2$, a small 4 will appear to the left of a period (.) on the head of the bolt. An 8 to the left of the period means the tensile strength is $80kg/mm^2$. Obviously, the higher the number, the stronger the bolt.

The number to the right of the period is ten times the ratio of the minimum yield point to the minimum tensile strength. It should suffice to say, the bigger the number, the stronger the bolt.

Those who are hung up on the use of American standards and want some way of approximating the ISO to the SAE can use the following chart:

SAE Grade	Property Class
Grade 2	5.8
Grade 5	8.8
Grade 7	9.8
Grade 8	10.9

Nuts

Like bolts, nuts are also graded by their strength. I have seen professional technicians spend a great deal of time locating the correct bolt for a given use, then mate it with any nut that is handy and fits. Common sense dictates the nut must be as important as the bolt, and reality dictates the use of a new nut during assembly. Now, I realize that every pro and home shop has a collection of used nuts and bolts that is cherished beyond reason. When rebuilding an engine, if you would like it to last at least as long as its first incarnation, use new nuts, and preferably new bolts as well.

Nuts should be replaced because they conform to the threads of the bolt. In doing so, they lose a little of their holding ability.

Washers

Washers come in several varieties, each with a different purpose. There are wave washers, split-ring washers, and star washers that are used to hold tension against the threads of a bolt or nut to prevent it from loosen-

Some bolts, such as those that hold cam towers in place, will have special machined shoulders to ensure the cam towers are properly aligned. Using the wrong bolts, just any bolt that seems to fit, can cause severe damage. For example, the cam tower shifts suddenly and slightly to the right at 3500rpm, the camshaft locks up, the chain or belt breaks, several valves come up and hit them, the valves break and are driven into the bottom of the cylinder head. This could be very embarrassing anywhere, but especially if it happened at high noon on the short cut between Lordsburg, New Mexico, and El Paso, Texas.

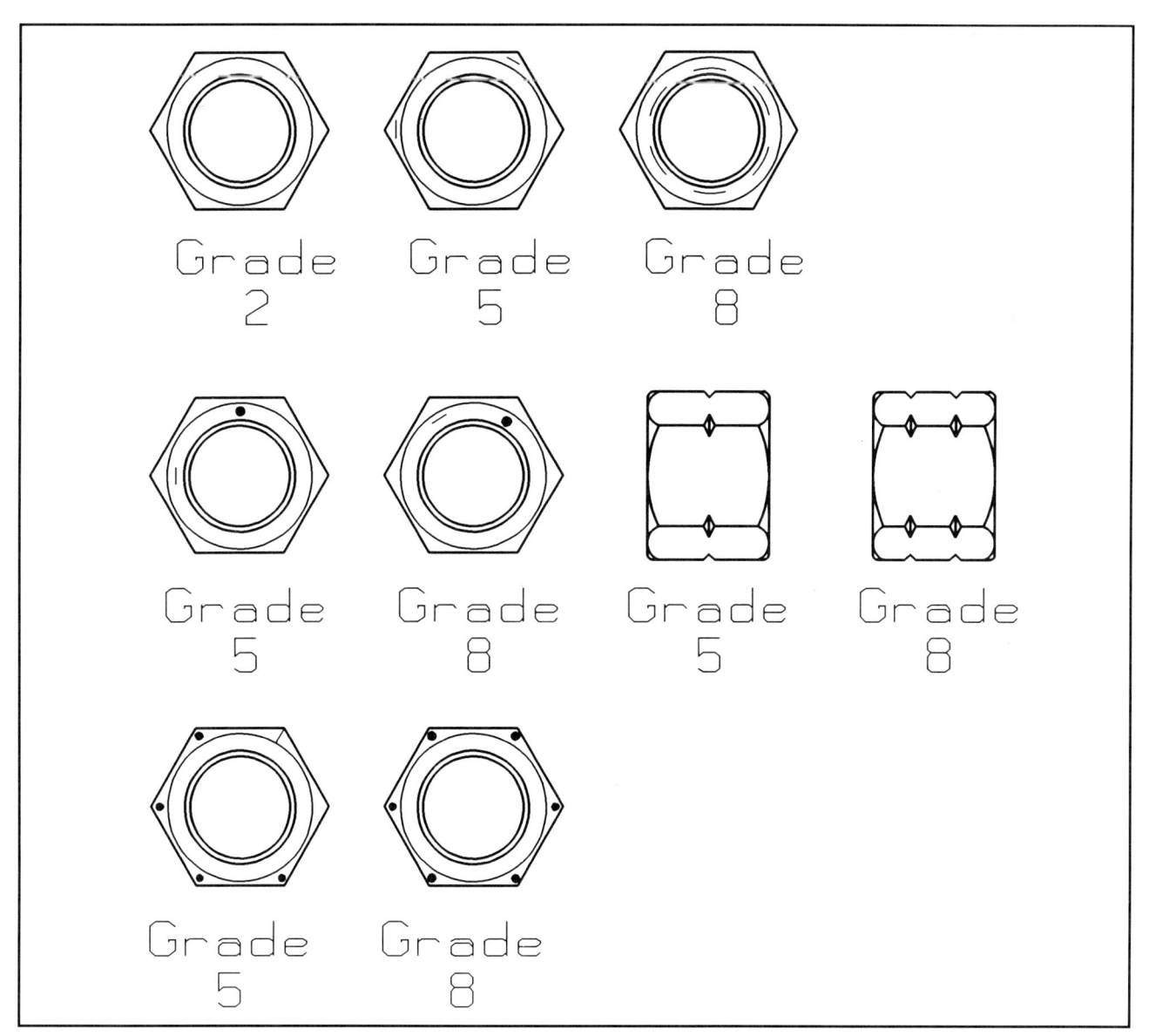

Like bolts, nuts are also graded on strength. Shown above are the meanings of three different marking patterns.

ing. These are often referred to by the generic name "lock washers."

There are two types of flat washers: through-hardened and case-hardened. The through-hardened washer is used to spread out the load of a tightened bolt across a flat surface. The case-hardened washer does the same; however, the core of the washer is soft and compresses under a load, allowing the bolt or nut to loosen. Approximately 30,000psi of clamping force is lost for every 0.001in the washer compresses. When selecting washers for use with a bolt or nut that has a torque specification, *always* use a through-hardened washer.

Studs

Studs are used in place of bolts in many places. A stud is threaded at both ends. Carburetors and fuel injection throttle bodies are often attached with studs. A few engines, such as the one found in early 1970s SAABs and in the Triumph TR7, use studs to hold the cylinder head in place. These should be replaced each time the head is removed. The disadvantage of using studs to hold the head in place is that they often corrode in place, making it extremely difficult to remove the cylinder head. In fact, I have often had to use a porta-power to remove a TR7 head. Unfortunately, this usually damages the head severely.

Studs should only be removed with a stud removal tool. Several tool companies market them.

Removing Broken Bolts

Rule number four in perfecting the art of bolt torquing goes as follows: *The likelihood of breaking a bolt is directly proportional to the difficulty in accessing the broken bolt.*

There are as many different types of broken bolt removal tools as there are surfboards on Oahu. All of them require drilling a hole in the center of the bolts, inserting the removal tool, and unscrewing the bolt. Before attempting to remove the broken bolt, apply oil around the threads. The removal tool is made of very strong, but brittle metal. If the threads are seized in the component, your first warning may be when the removal tool breaks. This will be an experience you will never forget. The risk of this occurring can be reduced by using the largest possible removal tool. Most of the bolt removal tools are harder than the drill bits you own. However, drilling may be your only alternative if the removal tool breaks.

Drilling Them Out

The alternative to the broken bolt removal tool is to drill the bolt out. Drilling can be done in two ways. One method is to drill the bolt to provide stress relief. Drilling a large enough hole in the bolt relieves the tension of the bolt threads against the threads of the hole. Often this, along with the largest possible removal tool, loosens the broken bolt sufficiently for removal.

If this does not sufficiently loosen the bolt, use increasing sized drill bits to enlarge the hole in the bolt until the threads can barely be seen. Use a chisel curl on what is left of the bolt.

Torching Them Out

If all else fails, you can use what the industry calls a blue-tipped wrench, an oxyacetylene torch. However, if you are not skilled with a torch to the extent that you can weld two drops of water together, pay an expert to do it.

Thread Inserts

Rule number five: *The likelihood of stripping a threaded hole increases as the degree of difficulty in even seeing that hole increases.*

There are several brands of threaded inserts to replace damaged or stripped threads. The kits

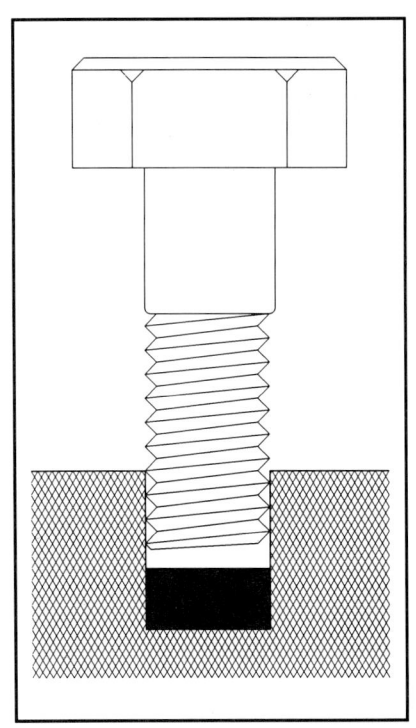

Damaged threads in bolt holes is one of the great curses laid on mankind by Synchro, the god of all things both automotive and frustrating. A relatively new product creates new threads from a resin. Pour the resin into the bolt hole and apply a releasing agent supplied with the kit to the bolt.

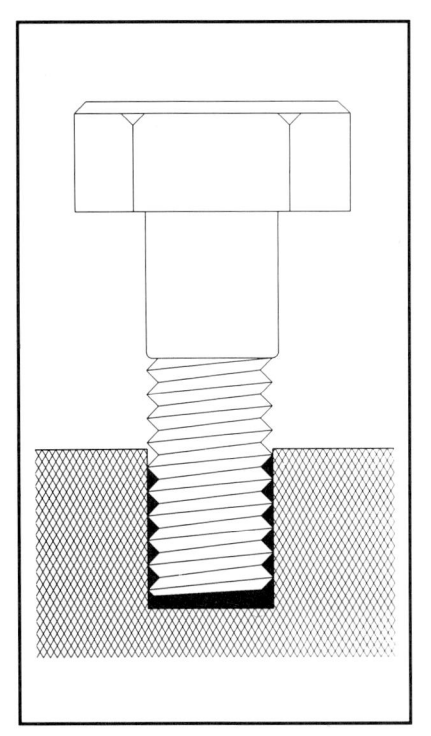

Carefully press the bolt into the resin. The liquid will form new threads around the bolt. Allow the belt to set for the length of time suggested by the manufacturer.

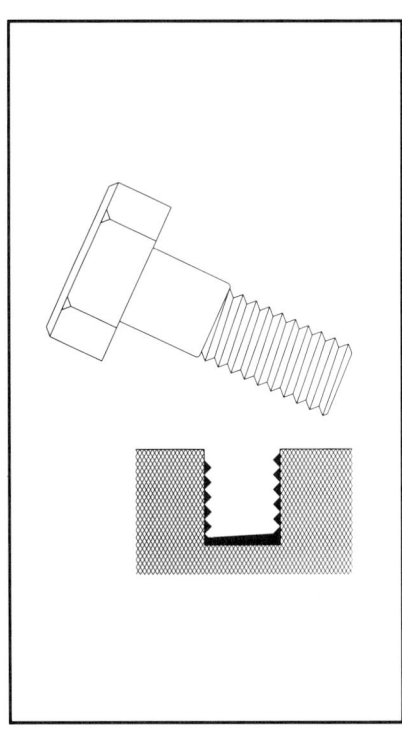

After the resin is set and the bolt is removed there will be new threads in the bolt hole. Unfortunately this technique does not work very well when the hole is in expensive metal. The art of torquing fasteners, Theorem 1 Corollary 25: *The chance of liquid threads working is inversely proportional to the cost of replacing the component that the threads are stripped out of.*

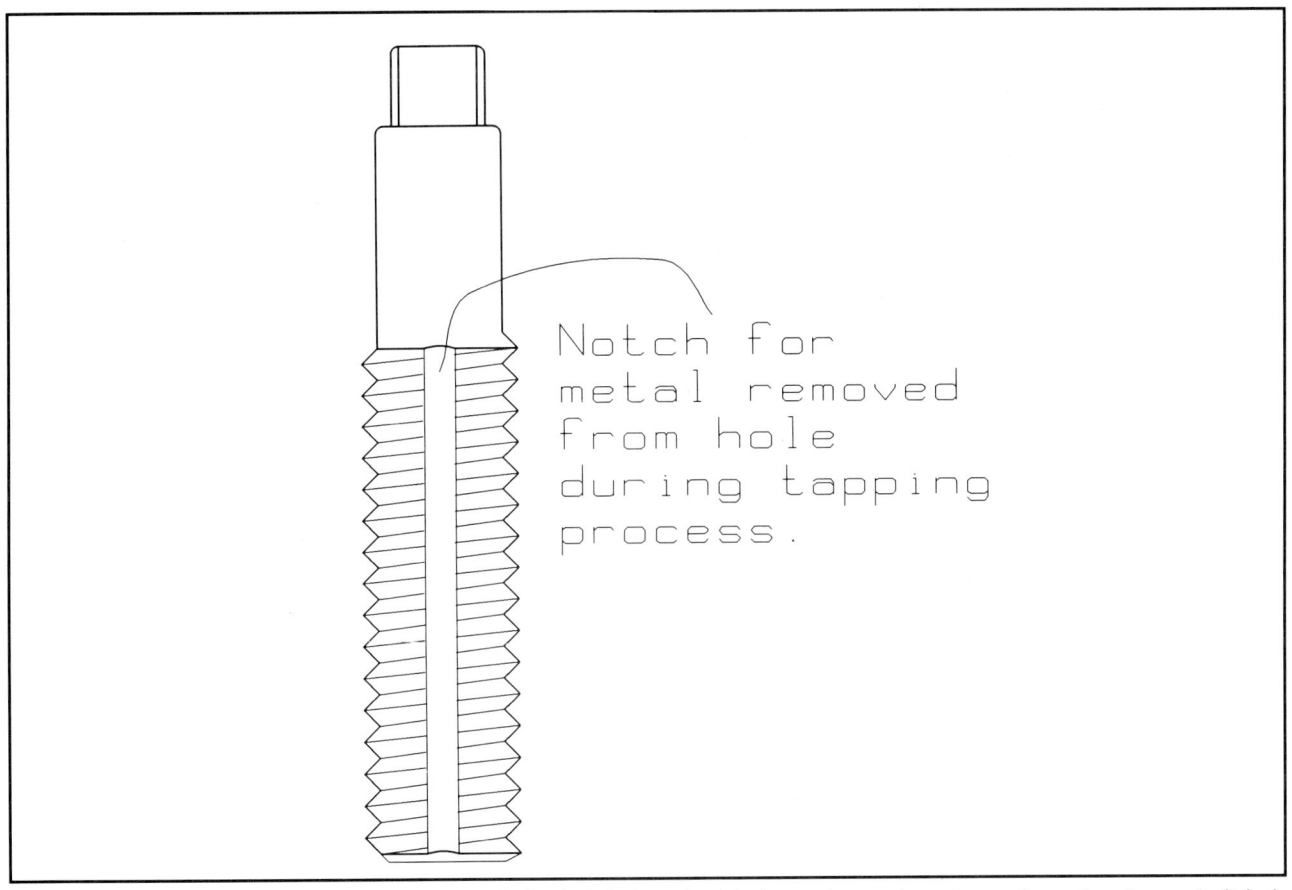

Notch for metal removed from hole during tapping process.

Taps are a very underused tool in most repair shops. When torque specs are critical, such as when torquing a head, the bolt holes should always be cleaned first using a tap. Also taps can be used to recut damaged threads or to cut new threads when a bolt hole has been redrilled oversize.

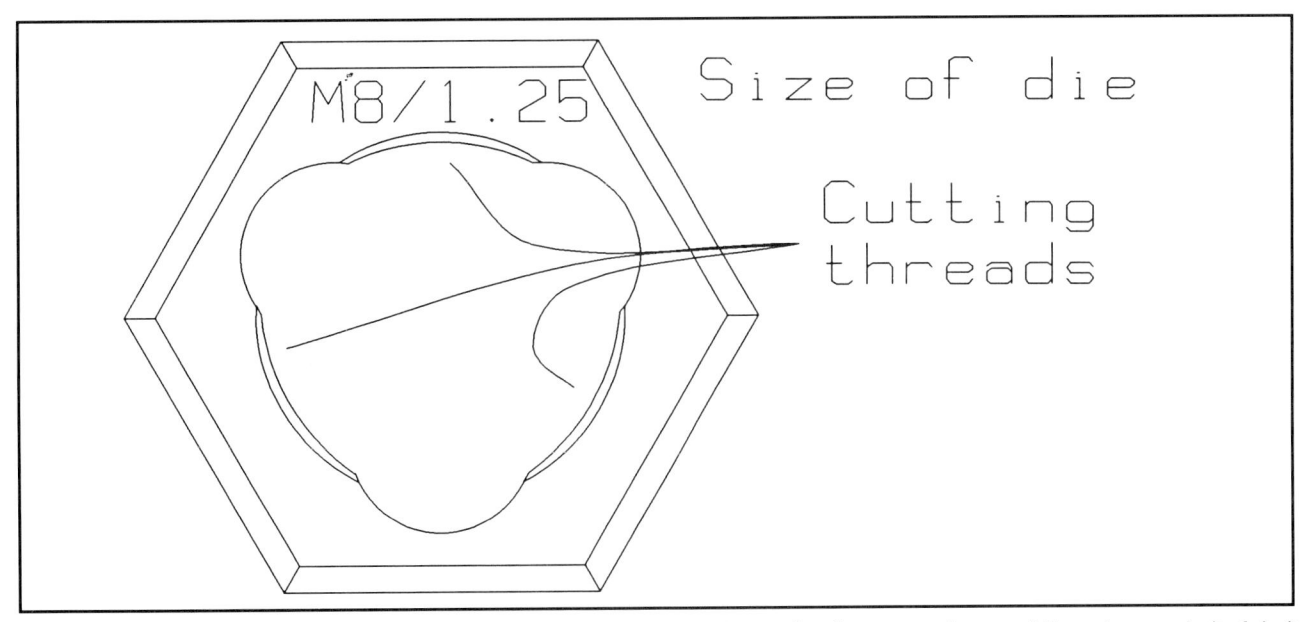

M8/1.25 — Size of die

Cutting threads

When I got my first set of Sears tools in trade school I wondered why it included a set of dies. An older "wiser" mechanic told me I would use dies far more often than taps. The reality is the most common use of a die is to clean up damaged threads on a bolt. A bolt with damaged threads should actually just be replaced.

33

Lined up neatly on paper shop towels is a set of head bolts. These should be lined up not-so-neatly in the trash can. Never re-use old head bolts. While you will almost always get away with it, while the guy down the street who had been a professional technician for thirty-two years may say he never did it, these bolts are stretched and should be replaced. Head bolts are a lot like men's underwear, and once it is stretched, it should not be used again.

for installing the inserts include a drill bit, a tap, a threaded insert installer, and threaded inserts. Drill the hole oversize with the bit supplied in the kit. Tap the hole and screw in the insert. Properly installed, these inserts have all of the strength of the original threads.

Liquid Threads

Rule number six: *What works great for everyone else will never work when it really needs to.*

One of the newer disaster relief products is a substance that could be called liquid threads. An epoxy glue is squirted into the damaged hole. A special releasing agent is then spread over the threads of the bolt. The bolt is then screwed into the hole and the epoxy is allowed to dry. After several hours, the bolt is removed. New epoxy threads replace the damaged metal threads. As rule number six states, however, when you *really* need this stuff to work...

Taps and the Art of Tapping

A better solution than the liquid threads, and in many cases better than the insert, is to drill and tap the hole. This is an option only if the situation permits a larger bolt to be used.

Begin by drilling the hole in question until all the old threads are removed. Be sure the hole being drilled is perpendicular to the mating surface, or the same angle as the original hole. A chart is provided in the appendix to help you choose the correct size drill bit for this task.

Lubricate the tap you have chosen with the appropriate type of lubricant for the type of material you are attempting to tap (see appendix chart). Slowly screw the tap into the hole a half turn, then remove the tap. Clean any metal shavings from the hole and restart the tap. Repeat this process slowly until the hole is fully tapped.

Dies

A die is a relatively useless piece of equipment. Actually, for every time I have used a tap I have used dies several times. Dies are useless in the sense that if the threads of a bolt or stud have been damaged enough to require their use, the fastener should be replaced.

If the bolt or stud is unusual and hard to replace, then a last resort would be to clean the threads with a die. Place the die over the end of the bolt or stud. Rotate the die a half turn and back it off. Like the tap, this helps to clean the metal shavings out of the new threads. Repeat this process until the bolt or stud is satisfactorily repaired.

Pipe Plugs

Pipe plug threaded holes are used primarily in the water jacket for fittings to channel water through the heater core and in the intake manifold for vacuum sources. On a given engine application if these sources are not being used these threaded holes will be plugged with a threaded plug.

Choosing Your Heads

5

Perhaps the easiest way to obtain cylinder heads that will improve the performance of an engine is to purchase them off-the-shelf. There are several manufacturers that market performance heads. Of course, this is the *easy* way, not the *fun* way. It is best to look for pristine heads, heads that have never been reworked, machined, or defiled in any manner since leaving the factory. Ask someone who is familiar with the type of engine and the type of performance work you are undertaking if there are different heads for that engine. For instance, the Ford 302 of the late 1960s and early 1970s had Windsor or Cleveland heads. The Cleveland heads were ideal for Granny to drive back and forth to the grocery store. The Windsor heads, on the other hand, from the 351 engine, were better suited for performance work. Many engines, especially those of later design and import engines, do not offer choices.

If you do have a choice, but have no luck in locating an engine expert to consult, here are a few tips to consider.

1. Choose the head that provides the most direct path for the air to flow into and the exhaust gases to flow out of the combustion chamber. The head with the fewest and shallowest angles probably has the greatest potential for airflow.

2. Choose the head with the biggest valves, or the one that has combustion chambers that will permit the modification for oversize valves. Seek the wisdom of a machinist who is experienced in such modifications. Just because it looks as though oversize valves can be installed does not mean they really can be.

Here lies the remains of a hundred dead engines. Most of the used cylinder heads you look at would more appropriately lie here than on the top of you street rod engine. Choose carefully, seek out tiny and hidden cracks lest your entire engine wind up here.

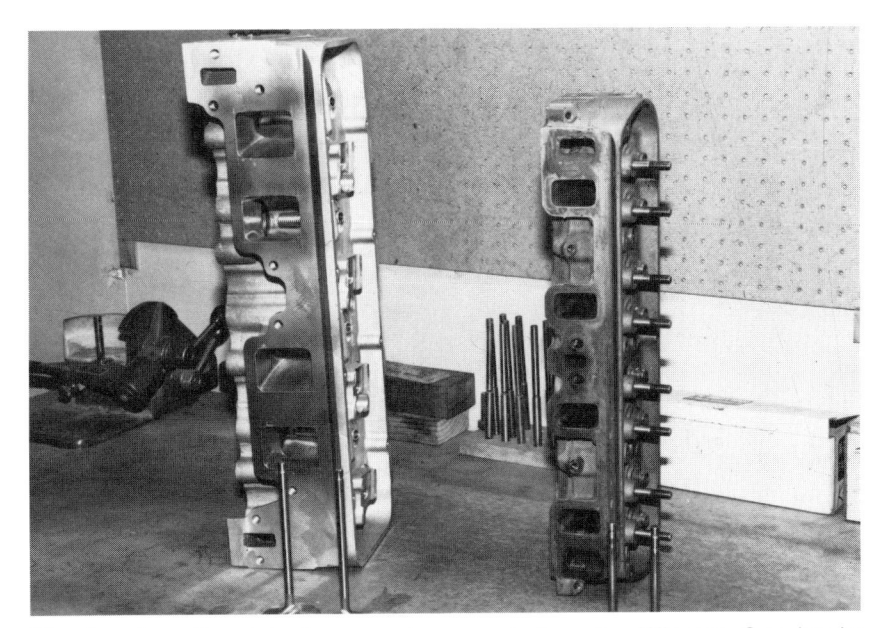

The head on the right is an early 1960s small-block Ford head. The one on the left was custom built for a Winston Cup car. Power comes from air, and since the Winston Cup intake ports are much larger than those on the Ford head, it obviously will allow the engine to gulp more air.

3. Choose the heads that have the best cross-flow arrangement for the intake and exhaust valves. In the typical cylinder head, the intake and exhaust valve lay in the same plane. This setup accommodates the pushrods and rocker arms. However, it severely limits the size of the valves and the combustion chamber cross-flow characteristics. Many times, I have been asked what is the advantage of an overhead cam configuration? My response typically has been that the main advantage is fewer moving parts or less inertial mass. The most important advantage is freedom from having to provide a straight path alignment between the camshaft and the valves to accommodate the pushrods. This virtually dictates that the valve be in the same plane. Pent-roof cylinder heads would be impractical without overhead cam design.

To summarize, the head should have a direct path for both intake and exhaust gases. It also either should have come with big valves, or have the ability to accommodate add-on oversize valves. The gas flow path through the combustion chamber should have the shallowest angles possible. Although there is no ideal cylinder head design, the best heads are the ideal balance between conflicting requirements.

Testing the Head

Before any head is used, even for a stock reconditioning, there are things that need to be tested. Look carefully for marks or other evidence that the head has been resurfaced. If the ultimate use of the engine is to power a grocery hauler, ask your machinist if any work that has been done previously on the head could affect the machine work he will need to do. If the head is to be used for performance work, it must be undefiled.

When satisfied that the head is suitable for your intended purpose, take it to your machinist and have him check it for cracks. Although unusual, even a new head can be cracked or otherwise damaged.

Pressure Testing

Perhaps the most important operation that can be done by the machine shop is testing the integrity of the water jacket in the block and head. Most machine shops are equipped to do pressure testing.

To begin the pressure test, all of the water ports in the head or block are plugged. Air or water is forced into the cylinder head or block. If air is being used, soapy water is sprayed onto the surfaces in question. If there is a leak, air bubbles will be evident.

Aluminum heads sometimes become porous. When this occurs, coolant literally will seep through the metal when pressurized. Should your machinist encounter this situation, he can sometimes seal it with a resin. Like so many other things in rebuilding an engine, because of the amount of

This head shows a nice direct air flow path from the intake manifold to the intake valve.

This is a counterflow head. Notice that the intake port of the head is positioned higher than the exhaust port of the head. Why do you suppose that is? Consider the gravity of that question.

work required to repair mistakes or bad judgment, a porous head should just be replaced.

Magnetic Crack Inspection

Magnetic crack inspection (also called magnafluxing) only works on ferrous metals, as it depends on the dispersion of the magnetic field through the metal. The principle is similar to the famous high school physics experiment that illustrates magnetic flux fields. In that experiment a magnet is placed under a sheet of paper and iron filings are sprinkled over it. As the iron filings fall, they line up along the magnetic fields of flux. The same basic technique is used in magnetic crack inspection. Iron filings are sprinkled on the head. A large electromagnet is then held a fixed distance above the head. Cracks in the cylinder head create distortions in the magnetic field, which affects the alignment of the iron filings.

Dye Crack Testing

Have you ever gone into one of those roadside diners along Route 66 in northern Arizona? Few people know that all of the cracked coffee cups in the universe are transported there by supreme forces beyond our comprehension. If you examine one of the cracks in the cup, you will find there is residue from the coffee of customers who sipped a little java while waiting for the bus to take them off to World War II.

Such a technique for head inspection is called dye crack testing. A penetrating dye is sprayed on the surface of the item being tested. After being allowed to soak in and dry, a cleaner is used to remove dye from the surface. Like the coffee stains in the crack of the cup, the dye will remain in the cracks of the component being tested. The dye method is especially effective on nonferrous metals such as aluminum and magnesium.

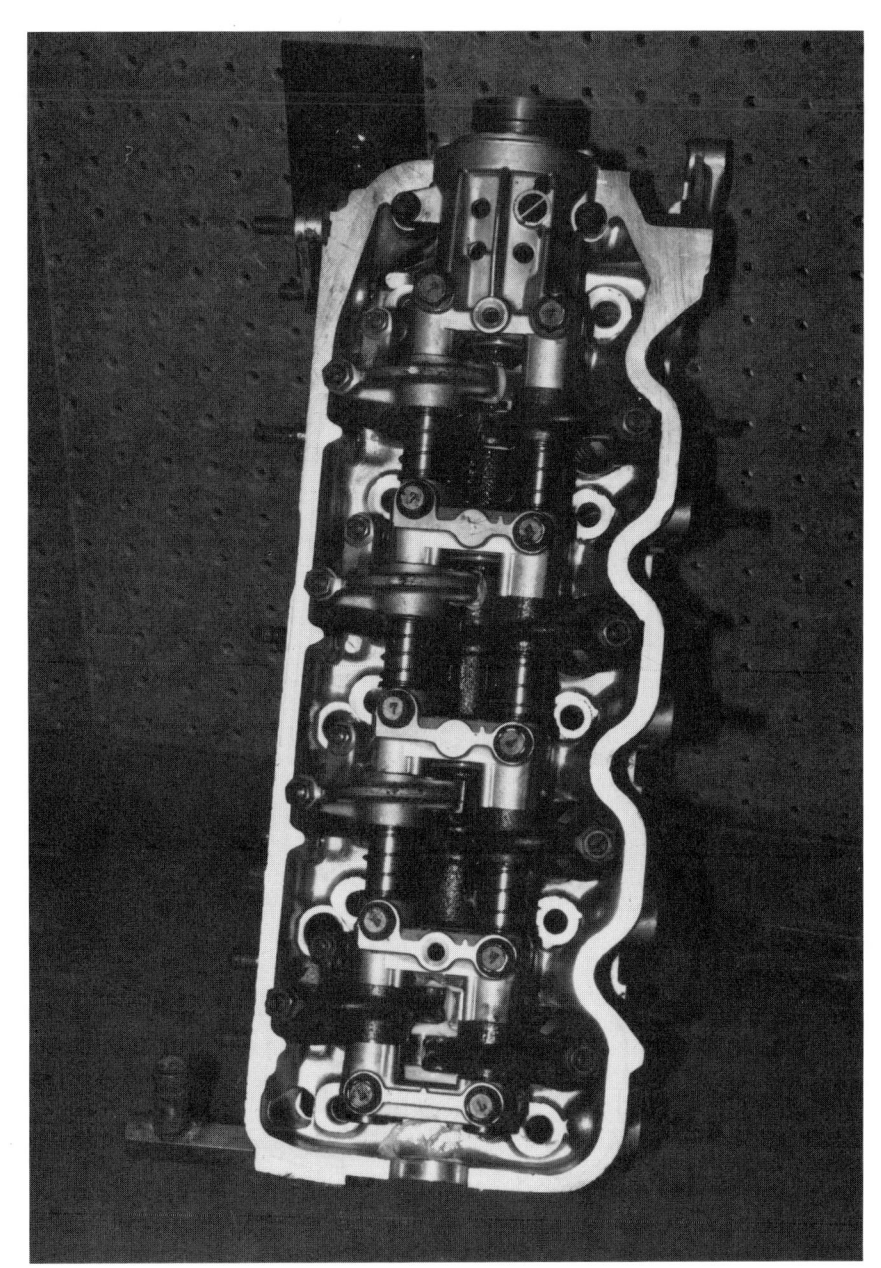

This is the head from the Honda CVCC engine. It is a cross-flow head with a semi-hemi combustion chamber. Like more modern engines it has a "multi-valve" design. There are two intake valves and one exhaust valve. Notice that the intake rocker arm has two adjusting screws. The one aligned along the centerline of the rocker arm is for the large main stratified (extra-lean) charge valve, the one off to the side is the small auxiliary (extra-rich) valve.

Cylinder Head Reconditioning

6

Resurfacing the Head

Of the machining operations done during an engine rebuild, resurfacing the cylinder head may be the most common. Constant changes of temperature coupled with inadequate maintenance such as routine retorquing of heads, or incorrect retorquing of heads, result in warping.

Resurfacing the head involves milling off several thousandths of an inch of metal. In effect this removes the high spots, making the head flat again.

Several things must be taken into account when the head is resurfaced. If the engine is an overhead cam design, removing several thousandths of an inch will cause the camshaft to be closer to the crankshaft. This can affect the tension or amount of slack in the timing chain or belt. Additionally, a warped bottom to the head may indicate that the top of the head is warped. If the top of the head is warped, the cam bearing alignment may be off. Misaligned cam bearings can bind, damage, and even break the camshaft.

Solutions? First of all, you need to realize that taking a link out of the chain to shorten it is not the answer. Within one or two revolutions of the crankshaft, the camshaft would be far enough out of phase to bend every valve in the head. As you read this you are probably thinking, "Oh, come on, no one would actually do that." Well, I once witnessed a professional technician in a hurry to finish a job doing that very thing. He knew better, he just was not thinking. So, think!

For many applications, automatic chain and belt adjusters will take up a great deal of this slack. When too much metal has to be removed, the slack can be corrected by shimming the cam bearing towers. This assumes that the application you are working on has removable cam towers. These shims can be obtained through your machinist or in some cases, through the appropriate dealer. For the applications where the cam towers are not removable, your machinist might be able to line bore the cam journals of the towers and install oversize cam bearings. There are head designs that do not lend themselves to this solution either, however, and they will simply have to be replaced.

Straightening the head may be another solution that eliminates many of the problems already addressed. The technology for this procedure is not available everywhere, and the degree of success seems to vary with the skill and knowledge of the technician. The process involves heating the head to a high temperature with it bolted to a machined flat plate. While this may seem simple enough to do in your oven at home, overheating the head can soften the metal or make it brittle, while not applying enough heat will be ineffective. Replace the head.

Misaligned cam bearings can be corrected by using the same procedure as described for tightening a chain or belt. Your machinist should be able to line bore the cam journals of the towers off-center and install oversize cam bearings.

Never consider using a cylinder head without first checking it for flatness. Spending a lot, or even a little, money for a head that turns out to be warped is foolish. The head will have to be resurfaced no matter what, however a warped head may force you to machine the head to a point where it increases the compression ratio to an undesired level.

38

Resurfacing the cylinder head is a relatively inexpensive procedure. Be sure that the machinist is aware of his limits in terms of compression ratio changes and how you are going to be using the engine.

If the engine is a V-design, both cylinder heads should be machined the same amount. Since resurfacing the cylinder head alters the size of the combustion chamber and affects the compression ratio, machining the heads differently will create unequal power and performance characteristics from each side of the engine. Additionally, removing metal from the heads can affect the interference angle between the intake manifold and its mating surface on the cylinder head. No more than 0.024in should be removed from these heads. If the head is badly warped, he will need to machine the mating surface of the cylinder head with the intake manifold. Not doing so creates several potential problems.

First, the ports of the intake

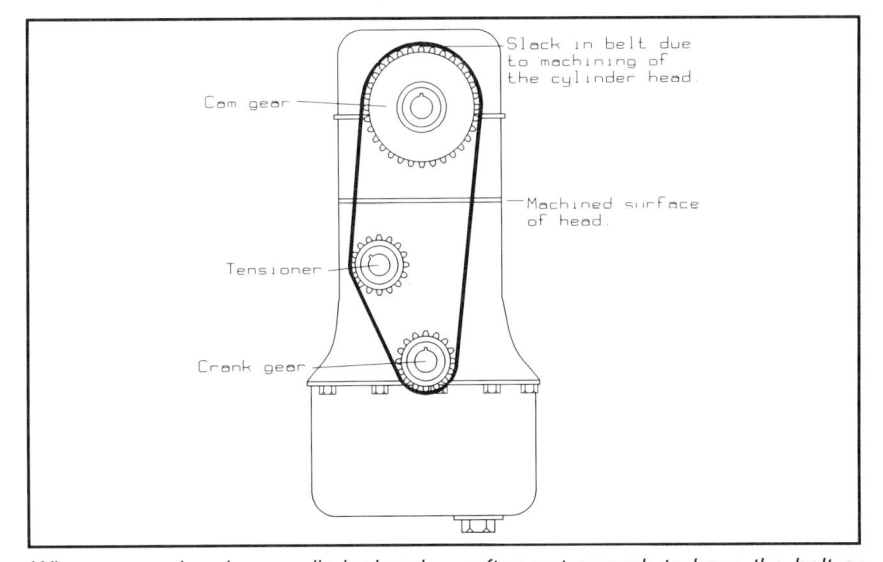

When an overhead cam cylinder head is resurfaced the camshaft ends up closer to the crankshaft. Belts and chains end up loose. Although most overhead cam engines are equipped with belt or chain tensioners, this is often not enough to keep the belt or chain tight. On heads with removable cam towers there are shims that can be installed under the towers to reduced the slack.

When a cylinder head is warped both the top of the head and the bottom of the head are distorted. When the head is resurfaced normally only the bottom is resurfaced. This can leave the cam journals out of alignment on overhead cam engines. For some heads the repair is a simple matter of machining the top of the head.

If machining the top of the head is not practical or possible, another alternative is to line bore the cam journals.

manifold may not line up properly. This can cause air leaks (often, and incorrectly referred to as vacuum leaks), oil leaks, and coolant leaks.

Second, since the rocker arm shaft or studs are closer to the camshaft, the rockers may bottom out the valve springs. Note that this problem is not unique to V-4s, V-6s, and V-8s. It can happen on any overhead valve design pushrod engine.

Third, removing metal from the mating surface of the head and block increases the compression ratio, which in turn increases the potential for detonation or pinging. Imagine after having spent $1,000 or more on repairs, and finding that the boss can hear you accelerate up the hill over five blocks away when you are late for work. The oil companies are shifting their emphasis away from anti-knock additives and toward emission control additives. Make every effort to retain the same compression ratio unless you are willing or desire to take some extraordinary steps to prevent spark knock. Raising the compression ratio may require the use of the most expensive gasoline available or the habitual use of a spark prevention additive.

The simplest solution to this problem is to use a specially designed shim or a thicker head gasket. Fel-Pro Gaskets makes a 0.020in shim available for several engines. Fel-Pro and other gasket companies also make head gaskets in various thicknesses. Check into these options with your local parts house and consult with your machinist. This solution solves all the potential problems.

Here is another option: If shims or thicker gaskets are not an alternative for the application you are working on, the intake manifold can be machined to ensure a perfect fit. This machining still leaves the problem of the shorter pushrods being required.

Refer to the following chart for how much to machine. Based on the head angle, and the amount machined from the head, the chart tells you how to determine how much to mill from the manifold-to-head-surface (For specific degree of head angle, multiply [amount milled] x N = amount to mill from manifold-to-head surface.)

Head Angle	Multiply Amount Milled from Head by:
5 degrees	1.1
10 degrees	1.2
15 degrees	1.4
20 degrees	1.7
25 degrees	2.0
30 degrees	3.0
35 degrees	4.0
40 degrees	8.0

From the manifold-to-block surface, remove (amount milled from head) x 1.71

Doing so still leaves the problem of the shorter pushrods being required.

For people born with a Hurst shifter "T-handle" in their right hand and who are impressed with flame-breathing monster iron, they do make shorter pushrods.

Valve guide knurling is another option. Although this should be done by someone with experience, kits can be purchased at a reasonable price to do it yourself.

Knurling the valve guides involves using a special tool to pull metal from the worn guide toward the center. This reduces the inside diameter of the valve guide to less than the diameter of the valve stem. A precision ream is then used to size the guide for the valve. While this may sound like a flaky repair, in reality this is an excellent way to repair the guides. Knurling leaves behind parallel grooves perpendicular to the centerline of the guide. These grooves help to control oil slipping down the guide to be burned in the combustion chamber.

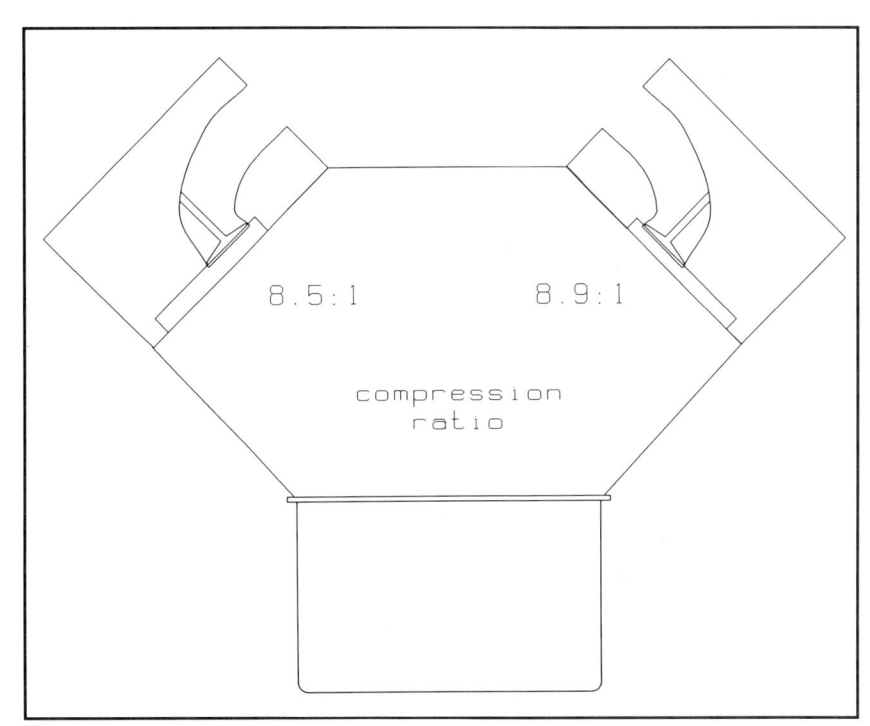

In the engine has two cylinder heads, a "V" or "pancake" design, both heads must be machined equally to ensure equal combustion chamber volumes. If the combustion chamber volumes are unequal between the heads, the compression ratio will be unequal between the heads.

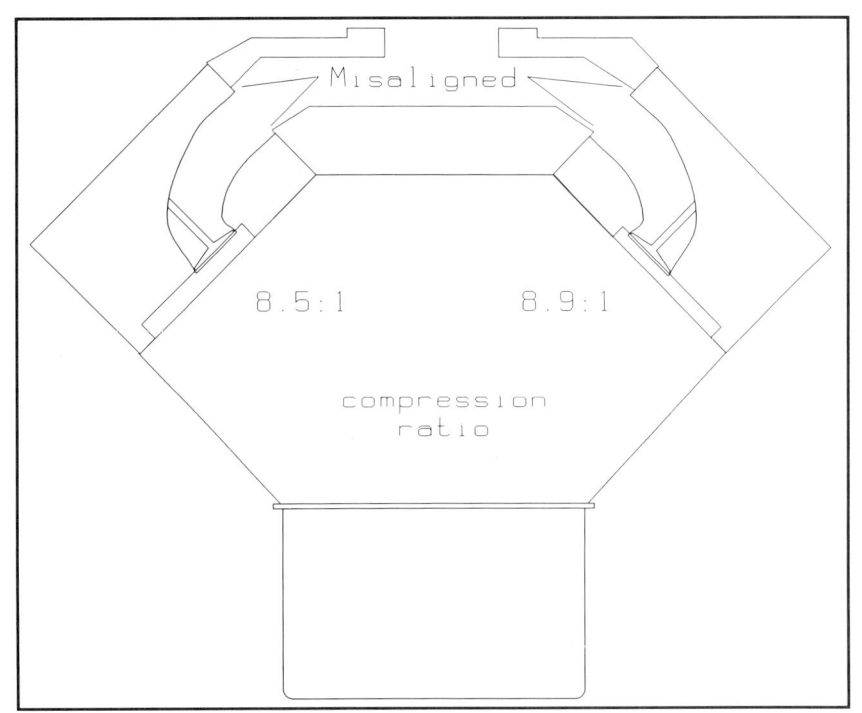

When the cylinder heads are machined on "V" engines, even if they are machined equally, the change in the distance between the bottom of the head and the bottom of the intake ports can cause a misalignment. The mating surface between the intake manifold and top of the block, front and rear, may have to be machined. This is done to the manifold, not the block.

41

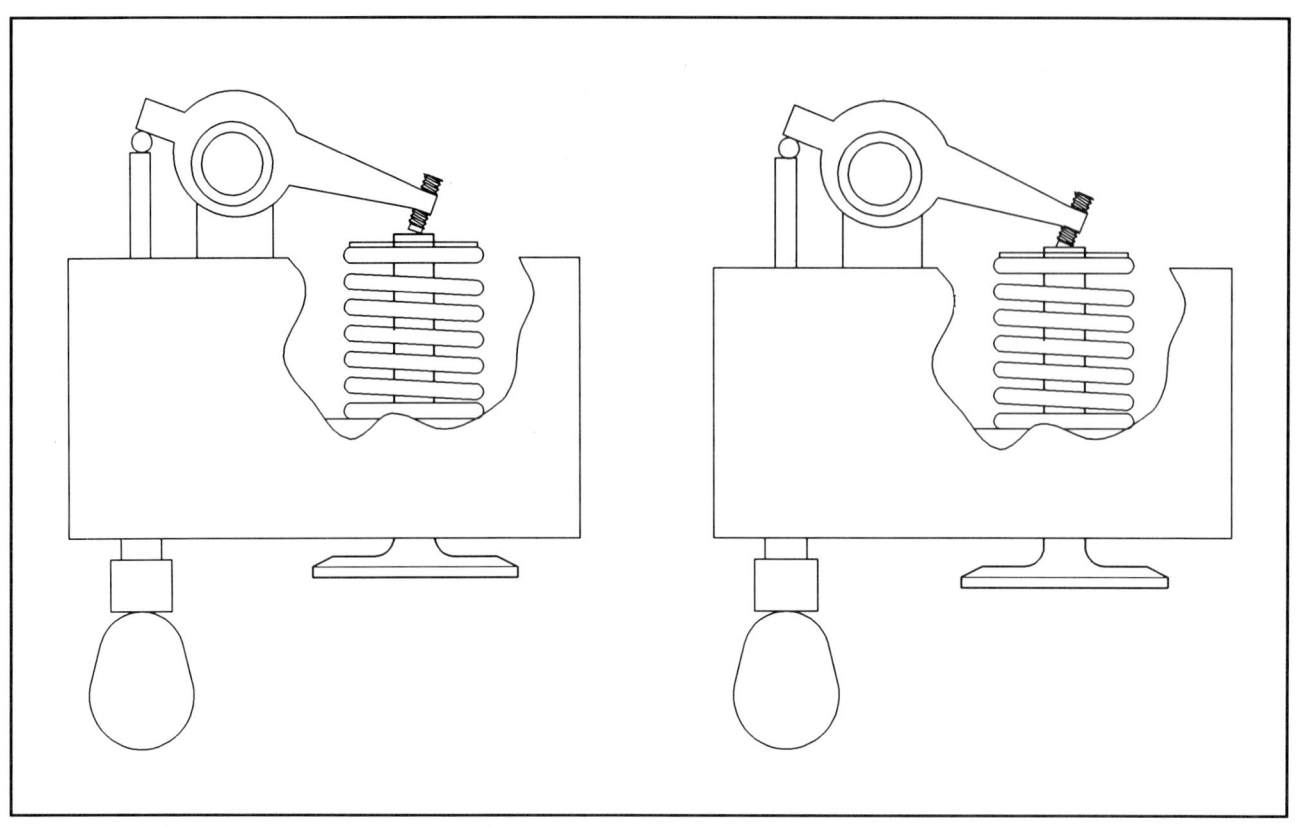

You probably thought I was done with potential problem associated with resurfacing cylinder heads. When the bottom of the head is resurfaced the rocker arm pivot moves closer to the camshaft. This may preclude proper adjustment of the valves or at least put the rockers at an odd angle, thereby increasing there wear factor. For many engines there are short pushrods available to alleviate the problem.

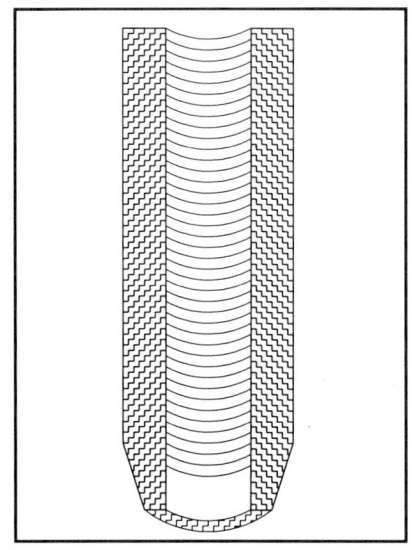

Knurling the valve guides is common practice on applications where the guide is part of the head. In this operation a tool is run through the guide to create grooves and pull metal toward the center of the guide. The guide is then reamed to the proper size.

If the guides are not replaceable, knurling makes good sense. However, many cylinder heads are designed with removable and replaceable guides. If they are removable, then replacement is the better option.

Replacing Valve Guides

Replacing the valve guides is a much easier operation than it might seem. The only special tools required are an appropriate valve guide drift and press or hammer. Slip the drift into the old guide and gently tap it out with a hammer. When the old guides are removed, install the new guide by tapping them into place. It is critical that the step of the valve guide drift be larger in diameter than the outside diameter of the guide. This is usually not a problem for drifts designed for this purpose. If the diameter of the drift is too small, it will damage the new guide.

Resurfacing Valve Faces

This process requires a valve grinding machine. At a cost of several thousand dollars, it is probably unreasonable to purchase a valve grinding machine for just one or two engine rebuilds. In addition, the procedure is one that requires skill and practice. Your machinist should always perform both this procedure and the grinding of the valve seats.

Resurfacing the valve face usually begins with chamfering the valve tip. This increases the accuracy of centering when the valve is chucked up in the arbor of the valve grinder. The next step is to insert the valve in the arbor of the grinder and adjust the arbor head to the correct angle. The face angle for most applications is 45 degrees.

The valve is held in the arbor and rotated at a relatively low speed. A grinding stone is also rotated at a relatively low speed. As a cooling oil is poured on the valve face, the machinist moves the arbor assembly across the stone. An adjustment wheel determines how much contact there is between the valve and the grinding stone. Metal is slowly removed from the face of the valve until it looks smooth and even all the way around. Any valve that shows a narrow band perpendicular to the valve margin during the first couple of passes is burned and should be replaced. If, when the face is smooth and even, an inspection of the margin reveals a sharp edge instead of a margin, replace the valve. Some machinists will cut a new margin. For a few dollars, it is better to have a valve that can be trusted for the next 150,000 miles rather than a valve that has already been in operation for 150,000 miles.

Based on that statement, you may be tempted to replace all the valves. Not a bad idea, but valves can be bent and damaged during transportation and storage. Even new valves should be refaced.

After each of the valves is refaced, the stem tips should be ground smooth. If the tip is badly worn, grinding it smooth will remove the hardened surface of the tip. This hardening is only a few thousandths of an inch thick. If in doubt about having passed through the hardening when grinding the tip, replace the valve.

If the valve neck accidentally comes in contact with the grinding stone, replace the valve. There is little doubt that this has weakened the valve in a critical area. Reusing it may cause the valve eventually to break off the stem and become impaled in the piston. While this can generate some unusual sculptures, it is hardly worth the trouble.

Note: Some applications use valves filled with metallic sodium. If enough metal is removed to expose the sodium to atmospheric moisture, it will *burst into flame!* Grinding these valves is extremely dangerous. I recommend that either you replace these valves instead of grinding them, or that you leave it to pro-

Quality valve grinding machines are very expensive. This Quick-Way machine has been the standard of the industry for valve resurfacing for decades. Newer machines offer digital displays and automatic controls but really do not do the job any better in the hands of a skilled technician than this old workhorse.

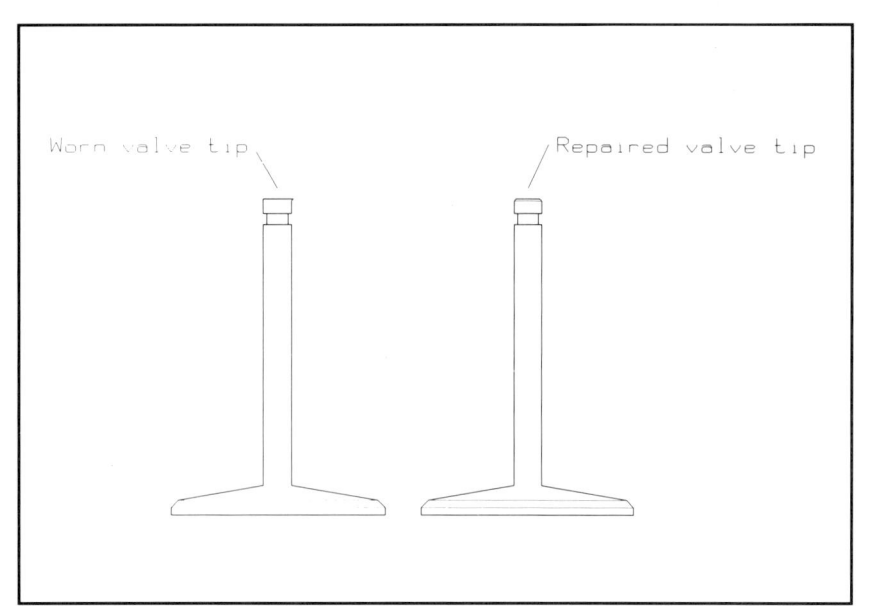

The valve tip should be cleaned up during the valve grind. The valve grinding machine will usually have provisions for machining and chamfering the tip of the valve. Actually, however, if the tip of the valve is that badly worn, and if the reader is building a performance engine, the valve should simply be replaced.

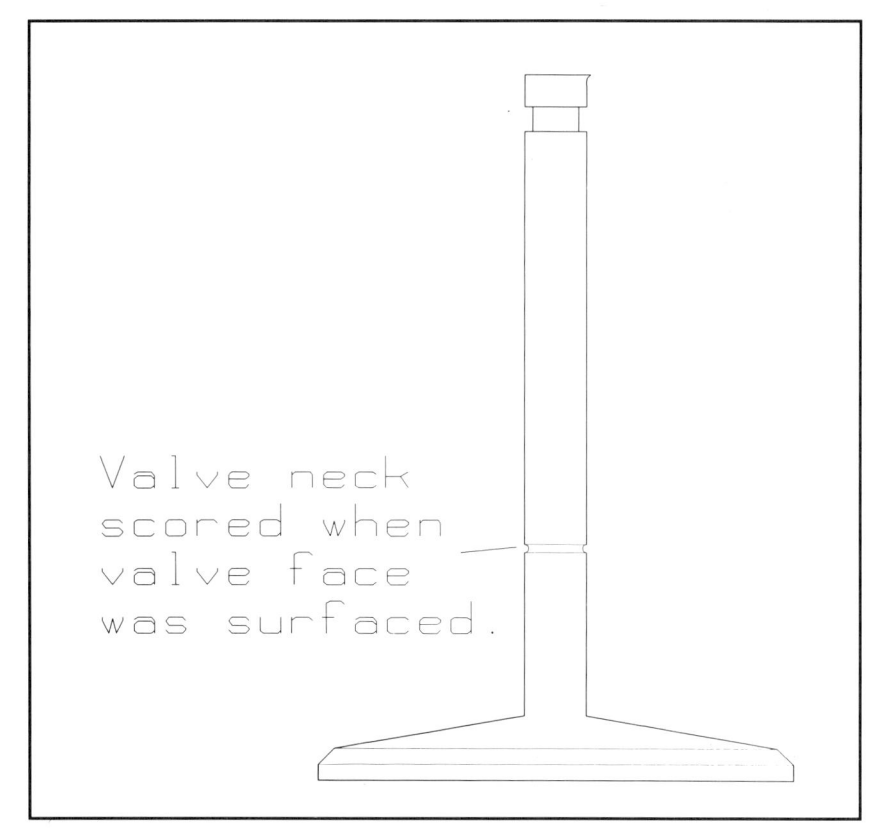

One of the common mistakes made during a valve grind is to allow the grinding stone to contact the valve neck while the face is being resurfaced. Should this happen, replace the valve. If you shrug and take a chance you may be replacing the cylinder head and a few pistons as well.

fessional, experienced machinists. If you are going to grind the valves yourself, and if you possibly are working with sodium-filled valves, contact your local fire department for tips on how to extinguish the fire in the event one does occur.

Resurfacing Valve Seats

Because the valve seat must be concentric to the guide, all valve guide repairs must be made before the valve seats are ground. There are three major categories of valve grinding equipment.

Back in the days of ducktails, white socks, and cuffs in your blue jeans, you might have found Billy Joe down at the town drive-in grinding his valves with a suction cup on a stick to impress the girls and lesser males. This method is known as lapping the valves. A spot of valve grinding or valve lapping compound is dabbed on the valve seat. This compound is an abrasive grit suspended in a lubricant. When the valve is placed on the seat and rotated back and forth rapidly, the valve and seat imperfections are theoretically matched. The reality is that matched imperfections are just imperfections. Lapping also does not adjust incorrect valve seat width. While valve lapping is valid when minor repairs are required, during an engine overhaul it is a wholly inadequate technique.

The second method of grinding valve seats requires a stone grinder. Generally, three stones are needed. The first stone, usually ground to a 45-degree angle, determines the seat angle. A second stone, usually ground to a 60-degree angle, narrows the seat from the bottom up. The third stone, usually ground to a 30-degree angle, narrows the seat from the top down. Correct seat width is important in the proper transfer of heat from the valve to the valve seat to the cylinder head. When no specifications are available, the intake valve seat width should be no more than 0.0625in, while the exhaust seat must be no less than 0.078in.

A machined pilot is inserted into the valve guide from the combustion chamber side. This pilot ensures that the centerline of the grinding stone is perpendicular to the ground surface of the valve seat. Next, the grinding stone holder is slipped over the pilot. A large electric motor, resembling a drill, called a driver is then lowered onto the stone holder. The driver is powered briefly, and removed from the stone holder while it is still spinning. Extreme care should be taken not to allow the weight of the driver to rest on the stone holder at any time. The weight of the driver can seriously affect the quality of the grind, as well as cause excess wear on the grinding stones.

After the initial 45-degree grind is made, the valve to be used in that seat should then be fitted in the seat. Place a spot of Prussian Blue on the valve face. Slip the valve into the guide. Using a lapping stick, rotate the valve against the seat at least twice. Remove the valve and inspect the valve face. The Prussian Blue should form a stripe roughly centered in the face of the valve. If the stripe is high on the valve, the machinist will use a 30-degree stone on the seat to lower the contact point between the valve and the seat. If the stripe is too low, he will use a 60-degree stone to raise the contact point. If the contact stripe is too wide, he will use both a 30- and a 60-degree to narrow the stripe. If the contact stripe is too narrow, he will use the 45-degree stone to widen it.

Many manufacturers and machinists will recommend grinding the seats to 44 degrees. The valve faces are usually ground to 45 degrees. This provides an interference angle of 1 degree. This slight, deliberate mismatch of angles is intended to help the valves seat. After as little as 100 miles, the interference angle disappears.

This high-speed grinding motor is used to resurface the valve seat. The cone shaped grinding stone provides one of the many angle that might be used to shape the valve seats.

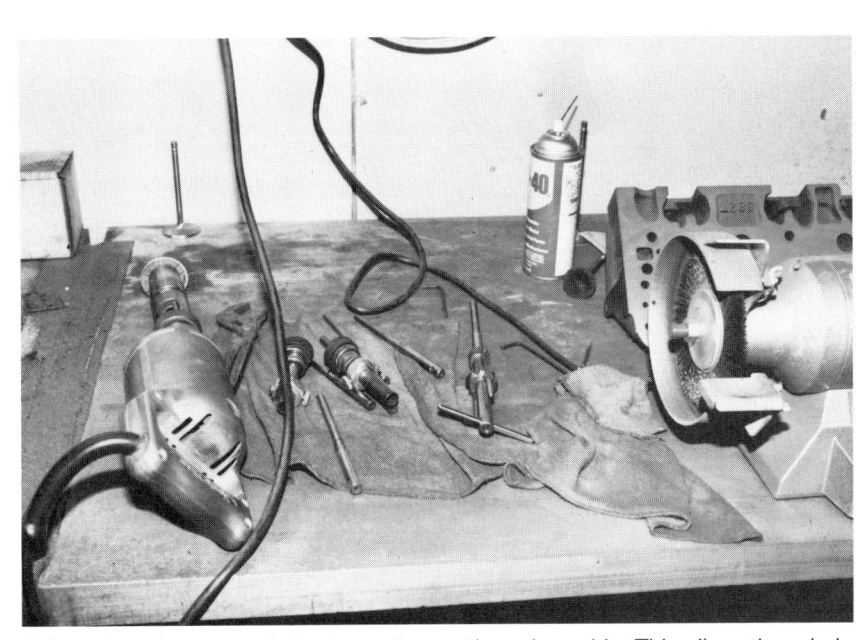

When the valve seat grinder is used a pilot shaft like the two seen in the center of the picture is slipped snugly into the valve guide. This aligns the grinder motor and the stone to ensure an accurate angle and a smooth grind.

Performance Valve Grinds

This is one of my favorite marketing terms. What if you do not buy the "performance" grind? Does the engine run slower and more safely? All valve grinds should be performance grinds. If the machinist insists there is a difference, do not argue, just pay the few dollars extra for the performance grind. Some shops, when doing a "standard" grind, will hit the seat with only a 45-

45

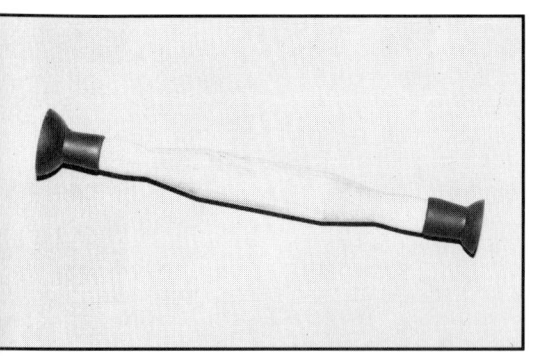

This is the first valve grinding machine I used. Imagine stone age man trying to start a fire by rolling a stick back and forth in his hands. That is how this tool is used, and that is the era in which this tool belongs. In fairness, it does have a place in finish work on the seats.

degree stone. This fails to ensure proper seat contact width or depth. Nevertheless, there is a distinction between the valve grind commonly sold as a performance valve grind and the precision valve grind that will be discussed later in this book.

Lapping Valves

Lapping valves never was an effective method of ensuring a seal between the valve and the seat. In the days of the Model T, this was often done when the carbon was scraped from the cylinder head. Even in those days it was an unsatisfactory method of grinding the valves.

Realigning Cam Bearing Journals

When the cylinder head has been warped on the bottom, it is probable that the head has also been warped on the top. This is a minor problem for engines that are not overhead cam. For overhead cam engines, however, this can cause a misalignment of the cam bearing towers. Misalignment can cause binding of the camshaft, which can cause the cam to break of the timing belt or chain to break. (*Note:* If you have had chronic problems with the timing belt breaking or stripping, it is a fairly safe bet that the cam towers are misaligned.)

The machinist has an easy cure for cam tower misalignment. The machinist will bolt the cam towers in place, line bore, and then install bearing inserts. The procedure is the same for applications where the cam towers are part of the head.

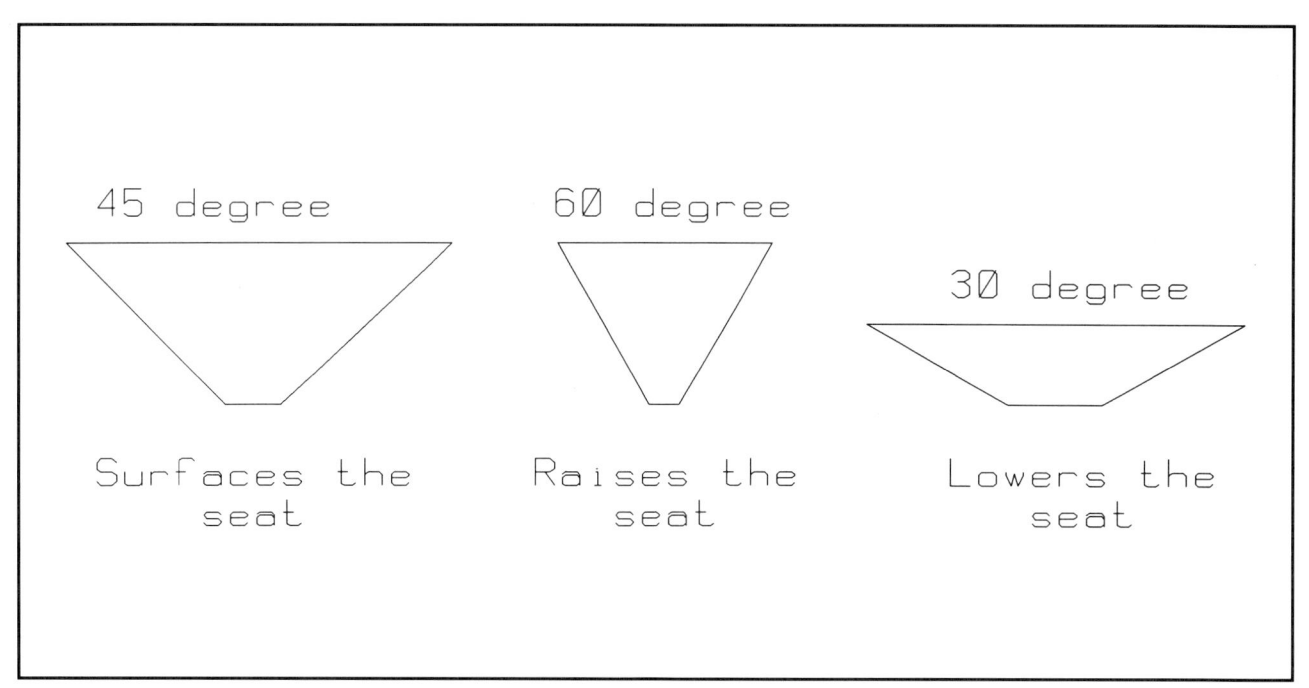

Three grinding stones are commonly used to form a good valve seat. The 45-degree (many technicians and machinists prefer a 44-degree angle) stone is used to form the contact surface between the valve and the seat. The 30-degree stone is then used to lower the top edge of the seat while the 60-degree angle is used to raise the bottom edge of the seat.

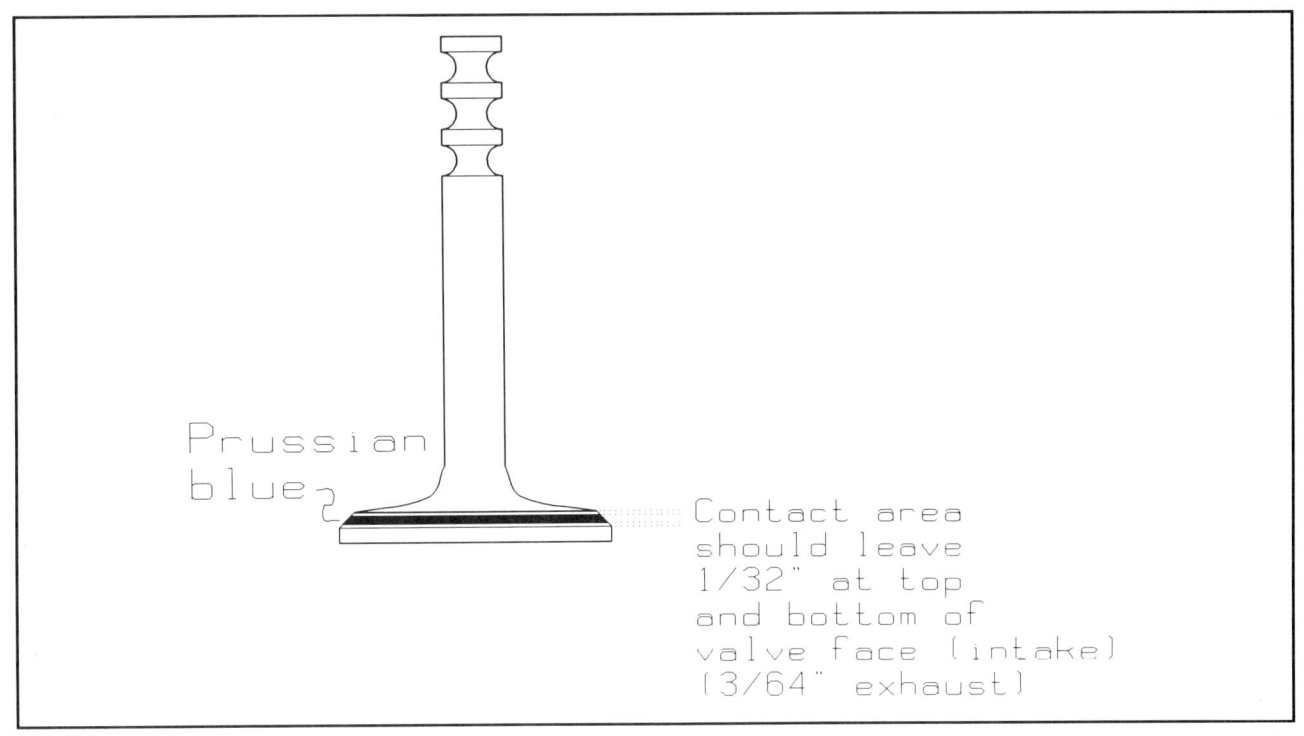

```
Prussian
blue                Contact area
                    should leave
                    1/32" at top
                    and bottom of
                    valve face (intake)
                    (3/64" exhaust)
```

Place a small daub of Prussian blue (you can get it at any real auto parts store) use the afore-maligned lapping stick to rotate the valve in the seat. The Prussian blue will leave a stripe around the face of the valve that corresponds to the valve's contact area with the valve seat.

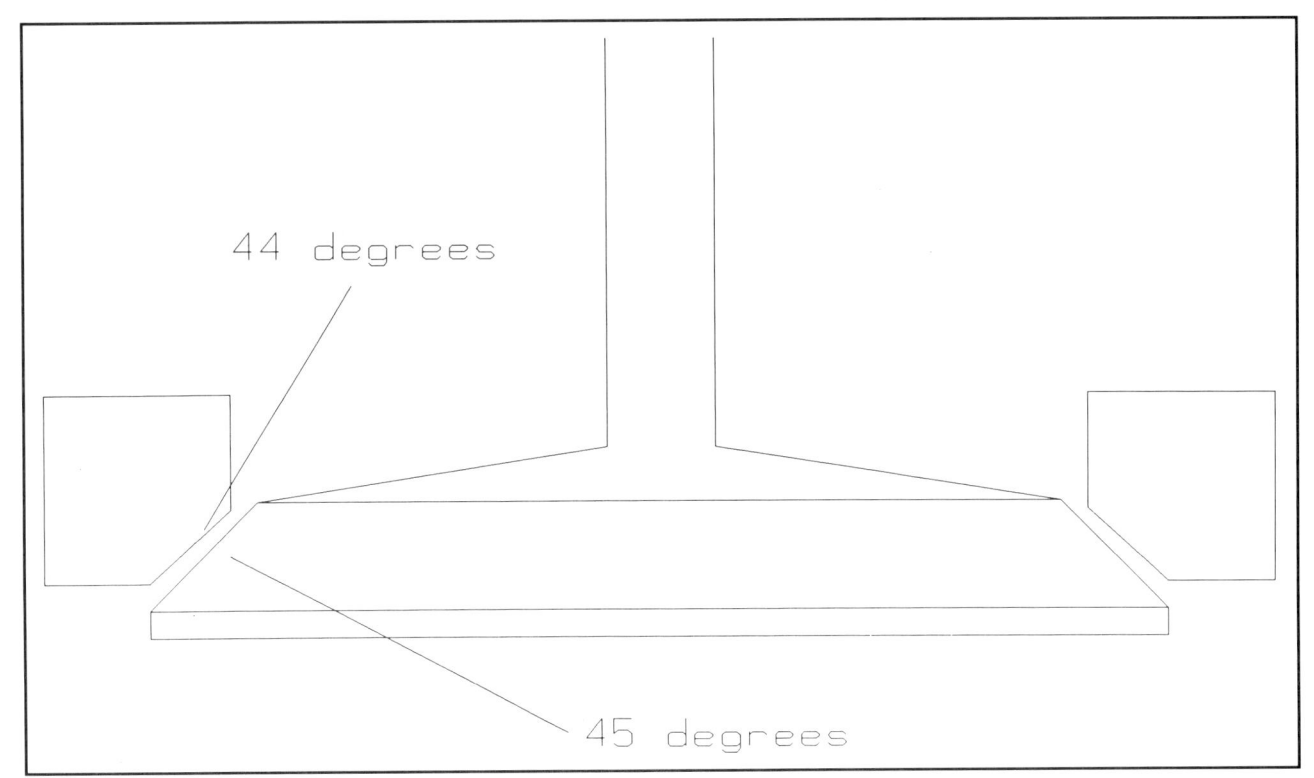

```
44 degrees



                    45 degrees
```

Many machinists will grind the valve seat at a 44-degree angle. When mated with a 45-degree valve face many feel that the valve will seat in sooner.

47

Airflow, Camshafts and Horsepower

To gain horsepower it is necessary to increase the mass of air in the combustion chamber. When heated, the expanding air drives the piston down the cylinder. In a gasoline engine, this process depends on mixing and heating the fuel and air into a combustible mixture. The mixture should then be spread evenly throughout the combustion chamber to take full advantage of the heat energy potential of the gasoline.

The energy in gasoline is measured in British Thermal Units (Btus). One gallon of gasoline contains approximately 130,000 Btus. At 100cfm of airflow into the engine (748.1gal or approximately 8.07lb of air) the engine requires 0.549lb of fuel (0.09gal) per minute of fuel. In one minute of operation, an engine operating at 100cfm can generate 11,700 Btus per minute, translating to about 276hp. This might seem like a lot of potential power—until the thermal efficiency of the engine is considered.

Thermal Efficiency

In a typical engine, 33 percent of the heat energy is lost to the exhaust gases, 30 percent to the coolant, 3 percent to overcoming friction, 3 percent to drive the fan and water pump, and 4 percent to pull air into the engine and push out the exhaust gases. With all of these losses, only 27 percent of the heat energy created during combustion is left to come out the back of the crankshaft. This means that of the 276 potential horsepower at 100cfm, less than 75hp make it out the crankshaft.

Volumetric Efficiency

Theoretically, a 302ci engine can move 437cfm of air at 5000rpm. Using the same calculations as for determining thermal efficiency, the engine would yield a potential of 328hp—assuming the cylinder heads can pass 437cfm of air through the valves at 5000rpm. Optimistically, stock cylinder heads provide a volumetric efficiency of 70 percent at 5000rpm, decreasing the cubic feet of airflow to 306. The poor volumetric efficiency reduces the potential horsepower to 230, equaling a rather sad 0.76hp per cubic inch. If the cylinder heads can be modified to allow 85 percent volumetric efficiency at 5000rpm the airflow will be increased to 371cfm, accompanied by a potential horsepower increase to 279. This yields a rather respectable 0.92hp per cubic inch. A further increase above 1.0hp per cubic inch could be accomplished by increasing the thermal efficiency of the engine. That topic, however, is outside the scope of this book.

The Camshaft

The cylinder head and its airflow capabilities are only as efficient as the valve control. The valves are opened and closed by the camshaft. Up to this point, our airflow calculations have assumed that the intake and exhaust valves are always open. Anyone who has ever been called

In a typical engine 33 percent of the heat energy generated through the combustion of gasoline is lost in the heat of the exhaust gases. Thirty percent is lost as heat energy to the coolant, 3 percent to friction, 3 percent to drive the fan and the water pump and 4 percent to the action of pulling air into the combustion chamber and pushing the exhaust gases out. This leaves only 27 percent of the power leaving the engine.

a motorhead knows that they are open only a percentage of the time.

For simplicity's sake, let's assume that the camshaft has a duration of 360 degrees on both the intake and exhaust, with zero time opening and closing ramps. Realistically this design is not possible, but it makes for an easier explanation easier. The 360 degrees means that both the intake and exhaust valve are open half the time, one full crankshaft rotation each. With a camshaft like this, the flow rate need only match the theoretical cfm of the engine. The square cam lobe would open the valve to its maximum at the beginning of the intake stroke. If the flow rate through the head intake port was exactly equal to the cfm at that engine speed, 100 percent volumetric efficiency would be achieved.

Unfortunately, camshafts cannot be built without opening and closing ramps. The valve both opens and closes rather slowly, which means that the airflow potential through both the intake and exhaust will be virtually zero when the valve begins to open, and will increase gradually as the valve opens. Fortunately, this gradual opening of the valves coincides with the piston velocity. The intake valve opens a few degrees before top dead center. As the crankshaft crests over TDC, the piston speed drops to zero momentarily. At this point the cylinder has no potential for drawing in air. As the crankshaft swings past top dead center, the piston begins to accelerate. As it accelerates, the potential for air to be drawn into the cylinder increases.

The piston speed and airflow reach their maximum as the crank journal for that piston swings past 90 degrees from TDC. After 90 degrees, the speed of the piston and potential airflow decrease. Fortunately, this decrease coincides with the closing of the valve as the lifter rides down the closing ramp. If the piston displaces 0.022cfm on each intake stroke, the bulk of the air to fill that cylinder must enter the cylinder while the piston velocity is at its greatest, between 45 degrees after TDC and 45 degrees before BDC. Eighty percent of the air must be drawn in during 50 percent of the intake stroke.

At 5000rpm, this engine would be demanding 436cfm of air. But most of the air must be drawn in during half the intake stroke, meaning that the airflow rate will be well in excess of 436cfm for half the intake stroke.

Volumetric efficiency affects the power potential of the engine.

Theoretical 328
Probable 230

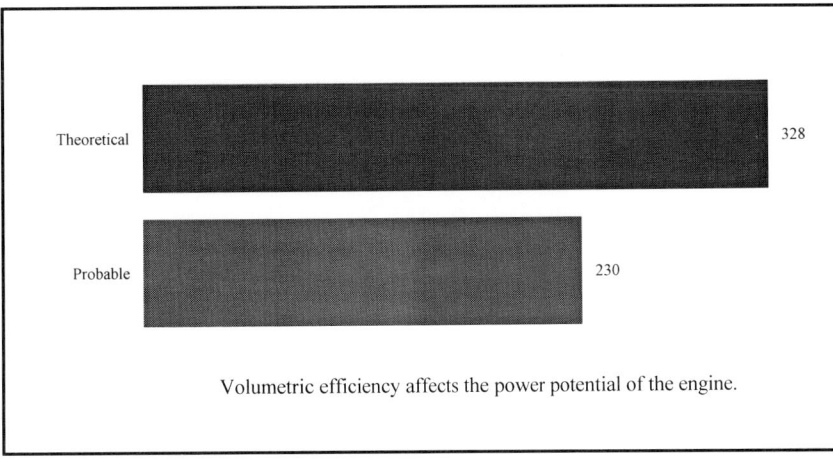

The power generated by an engine is directly related to the amount of air that can be drawn in. In the above example the engine is capable of drawing in 328cfm at a given rpm. However, there are restrictions in the intake system, the exhaust system, and most importantly for this publication, the cylinder head intake and exhaust ports. Also the flow path through the cylinder head (cross-flow, counterflow, etc.) can affect volumetric efficiency. Some stock engines, such as the GM Quad4 approach 100-percent volumetric efficiency in certain rpm ranges.

The camshaft is as important to the flow of air into the engine as the design and configuration of the intake or cylinder heads.

49

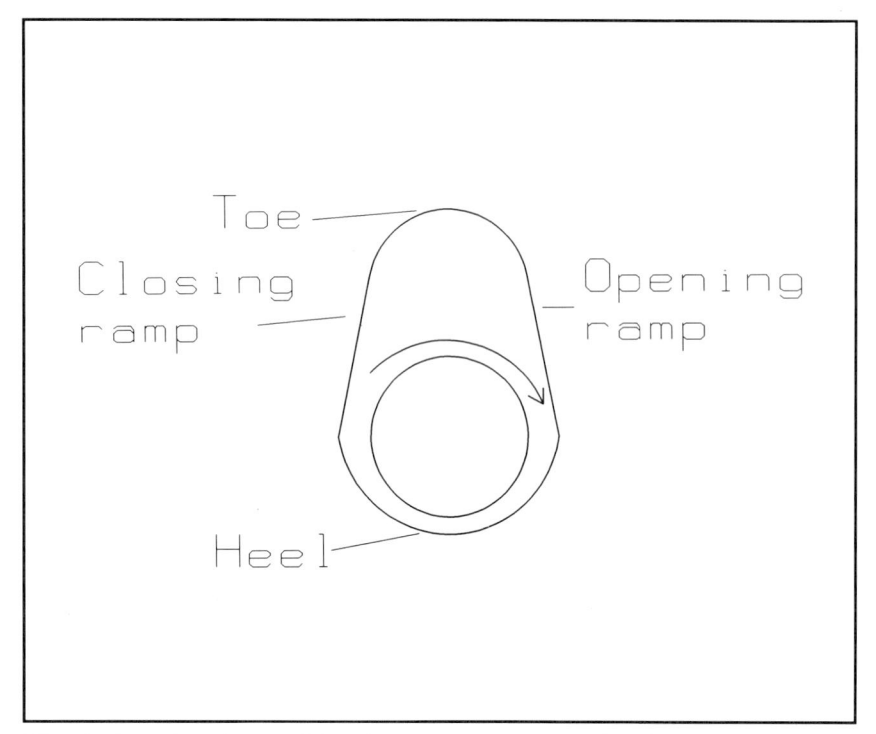

The design of each cam lobe determines how fast the intake valve and the exhaust valve will open and close. When the lifter is on the heel of the cam lobe the valve is closed. As the cam rotates the lifter or follower rides up the opening ramp, across the toe where the valve is at maximum lift and then down the closing ramp.

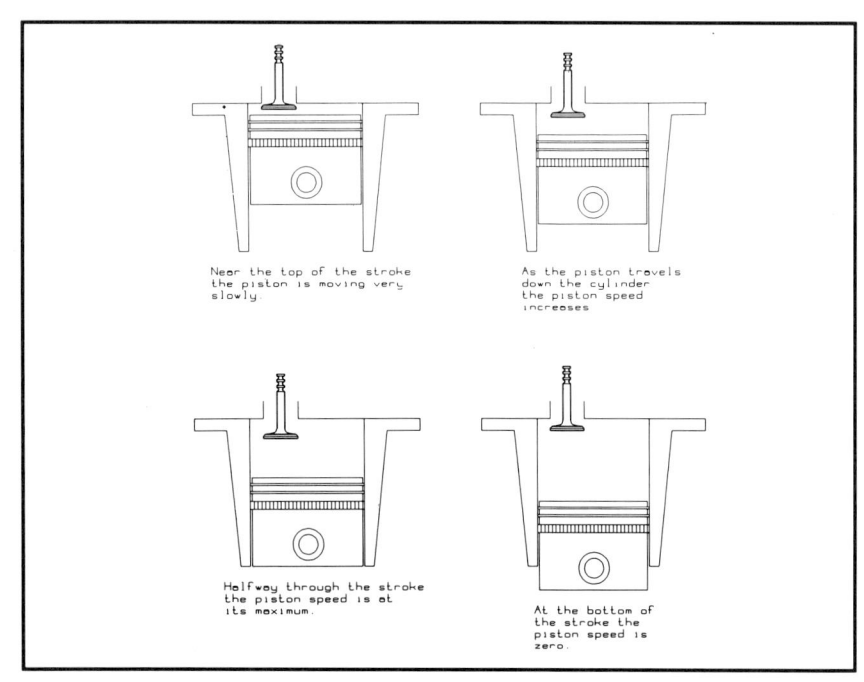

The movement of the valve is controlled by the camshaft. The phasing of the camshaft to the crankshaft is critical. Camshaft manufacturers have recommended cam timing specifications for maximum power and effectiveness. The camshaft must be accurately timed, or degreed, to maximize its effectiveness.

The heads, therefore, must be able to flow 654cfm, which is equal to 81.75cfm per cylinder. In fact, this flow rate can marginally support engine rpms approaching 7000.

[(Cubic inch displacement/ 1,728) x RPM] / 2 = CFM

CFM x 1.5 = Required Head Flow

Keep in mind that the formula shows required airflow for *all* cylinders. To determine the amount of airflow per cylinder, divide the result by the number of cylinders.

Up to this point, in our examples we have worked with a maximum of 5000rpm. This was a convenient number to work with, and is a good maximum rpm for a street engine. But what factors, other than use, determine the maximum rpm of an engine?

Although there are many moving parts in an engine, none are subjected to more stress than the piston. A stock engine, with stock pistons, rods, and crankshaft, will have a maximum piston velocity of 3,500 feet per minute (fpm). The maximum rpm on a 302 with this piston speed is 7000. Apparently, the figures used up to this point are very conservative. Full-race forged cranks, heavy-duty rods, and main bearing caps and pistons can support piston velocities of 6000fpm. With this type of lower end preparation, rpms approaching 12,000 are theoretically possible.

RPM = (FPM x 6)/Stroke

For the 302ci engine, 7000 rpm would require a per-cylinder airflow rate of 114cfm.

Selecting a Camshaft

While the cylinder head is being flow tested, the technician will measure airflow at various lift points for the valve. The camshaft should be able to lift the valve to the point where the desired airflow can be achieved. A camshaft with a lift of 0.500in will open the valve 1/2in. If

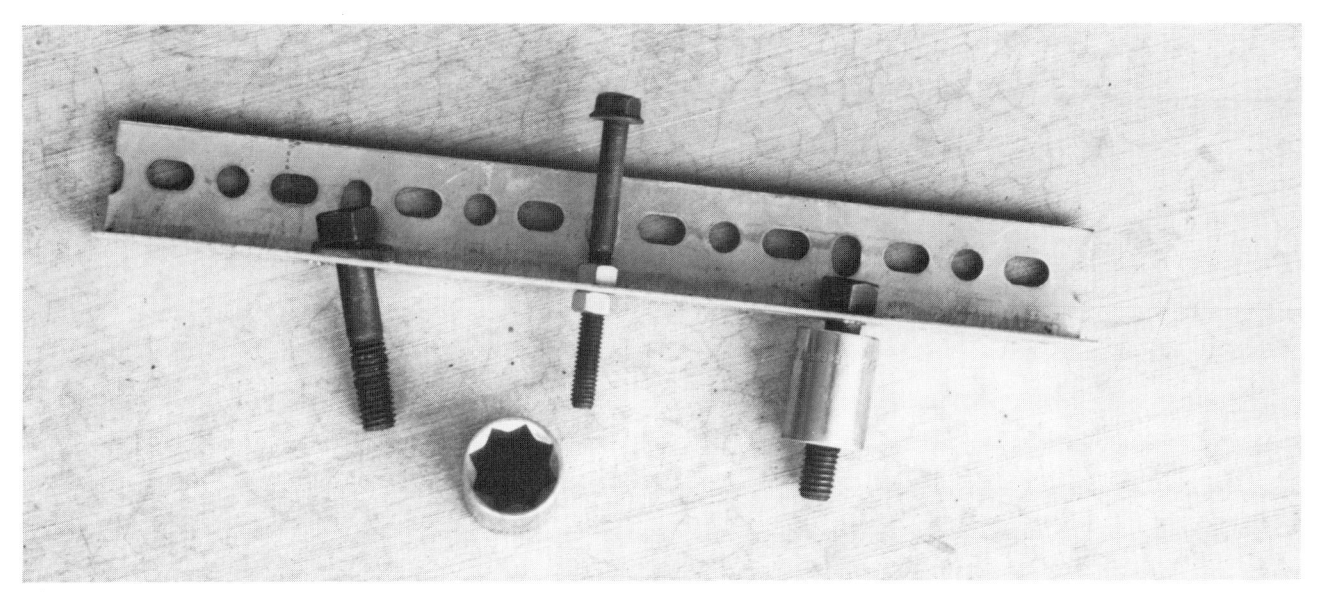

To degree the camshaft it is first necessary to identify the exact point of TDC for the #1 cylinder on the compression stroke. A piston stop is required to do this job. Piston stops can be purchased commercially or can be manufactured from a piece of angle iron and a few bolts. This "homemade" piston stop uses the old head bolts that I recommended throwing away in a previous chapter.

Install a degree wheel on the front of the crankshaft. Position the degree wheel so it is near the top dead center mark when the piston appears to be near top dead center. The adjustment is not critical at this point. Here a pointer has been made from one of the most useful supplies found in any shop environment: an old coat hanger.

51

Set the piston stop so the piston will only travel to within 1/4in–1/2in of the top of its travel. Rotate the crankshaft clockwise until the piston comes against the stop. Note the reading on the degree wheel. Rotate the crankshaft counter-clockwise until the piston comes against the stop. Note the reading on the degree wheel. Top dead center is halfway in between.

Bring the crankshaft back to TDC and readjust the degree wheel to read zero.

114cfm is our target, then the head must flow a minimum of 114cfm when the valve is opened 0.500in. On many overhead cam engines that do not incorporate a rocker arm, the formula for cam lift versus valve lift is just that simple. However, for engines that use rocker arms, the ratio of the rocker arm must be considered.

Cam Lift, Valve Lift, and Rocker Arm Ratio

Cam lift is the difference between the largest diameter and the smallest diameter of the cam lobe. Camshaft manufacturers advertise lift measured in inches of gross valve opening, assuming a stock rocker arm ratio.

If the rocker arm ratio is 1.5, it means that a 0.500in movement of the pushrod will result in a 0.750in movement of the valve. In the example, the target airflow was 114cfm. After having the finished heads flow tested, it is determined that this airflow rate is achieved at 0.500in valve lift. The camshaft could have a 0.500in lift, or a 0.450in lift could be used with a 1.5:1 rocker. This combination would provide a valve lift of 0.675in. The result is more than adequate airflow without the extremely high stress factors that would be placed on a high-lift camshaft. A rocker ratio of 1.5:1 is typical for a stock rocker arm. Some performance rocker arms are ratioed at 1.7:1. Using the same example, the 1.7:1 rocker would increase the valve lift to 0.850in. This example is pretty radical, though, and would probably either result in the valve spring retainer bottoming on the valve guide, or the spring coils binding.

Net lift (valve opening) is equal to the cam lift, less camshaft flex, times the rocker ratio, less valve clearance and valvetrain flex. Therefore, a cam of 0.300in, less cam flex of 0.010in, would yield a net cam lift of 0.290in. A 1.5:1 ratio rocker would yield a gross valve lift of

Install a dial indicator on the intake lifter and rotate the crankshaft against normal rotation until the dial indicator reads its most extended reading. Now rotate the crankshaft with normal crankshaft rotation. When the valve lift reaches the point recommended by the camshaft manufacturer the degrees shown on the degree wheel should also match the readings recommended by the manufacturer.

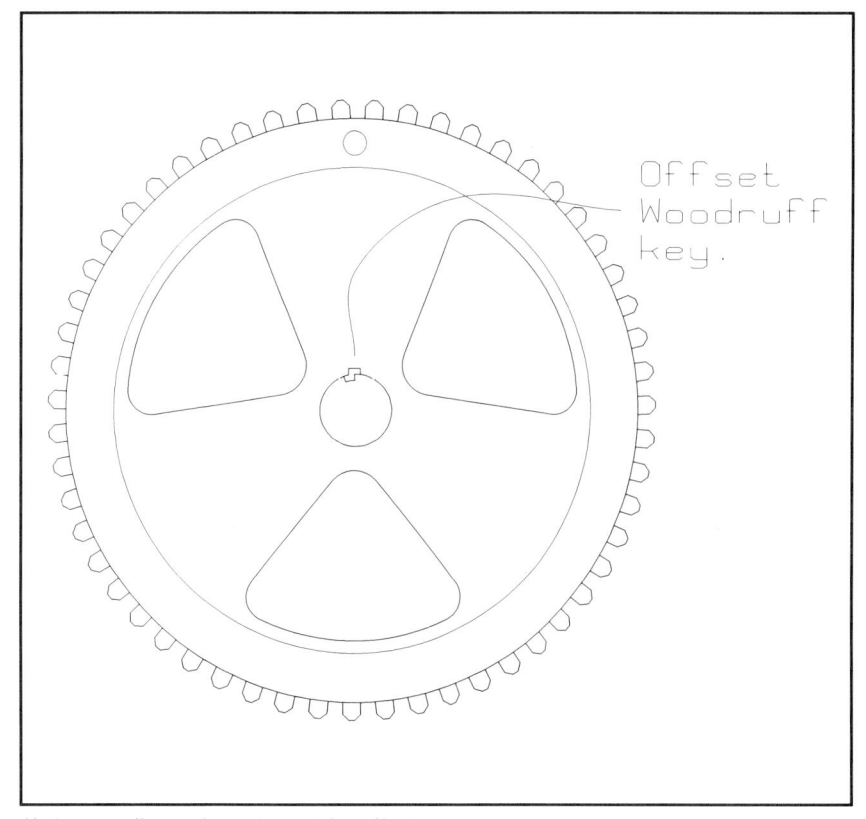

If the readings do not match, offset keys can be purchased to make it correct.

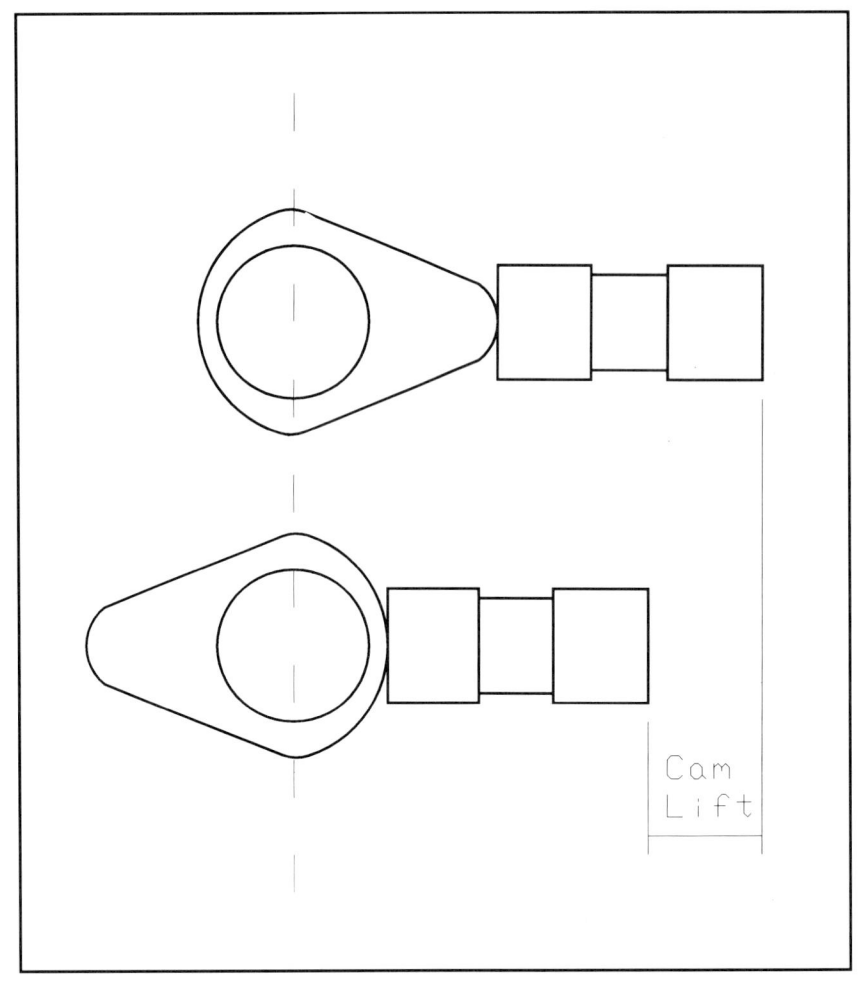

Cam lift is the difference between the rotational centerline of the camshaft and the cam lobe toe, minus the difference between the rotational centerline of the camshaft and the cam lobe heel. In other words, the amount the cam lifts the lifter or cam follower (in the case of overhead cam engines.)

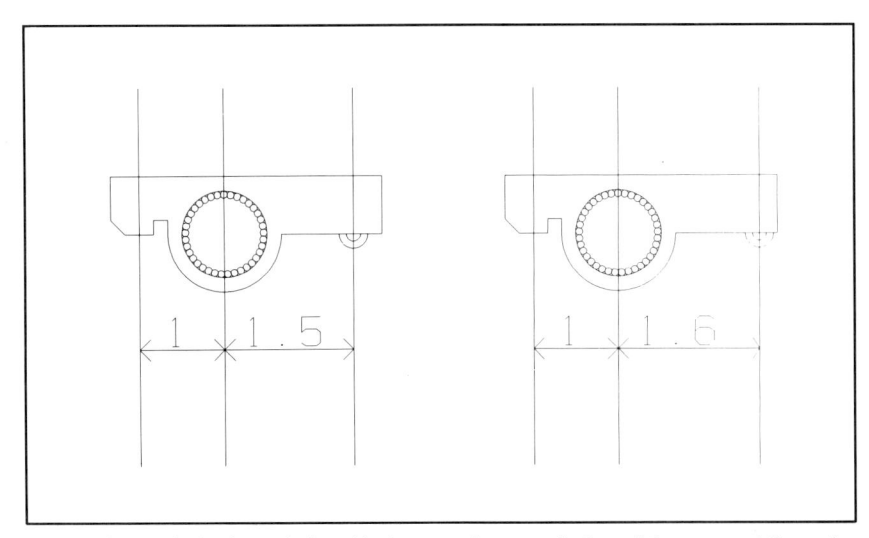

The rocker ratio is the relationship between the movement of the rocker by the camshaft and the amount the valve is moved by the rocker.

0.435in. If the valvetrain flex (flexing of the pushrod, for instance) amounts to 0.004in and the valve clearance is 0.018in, the net valve opening will be 0.413in.

Now we have arrived at a quandary: Do we select the camshaft and rocker arms based on predicted cylinder head airflow? Or do we maximize the efficiency of the heads, have them flow tested, then select the camshaft and rockers? The obvious conclusion is that the flow rate through the cylinder head is the variable over which the engine builder has the least control. Maximize the cylinder head flow, have the head flow-bench tested, then select the camshaft lift and rocker ratio.

Duration

Camshaft duration is measured between the valve being opened a specified amount and that same amount prior to closing. The Society of Automotive Engineers (SAE) standard for measuring duration is at 0.006in of valve lift, while the performance industry measures at 0.050in valve lift. This means that one must be careful to make apples-to-apples comparisons when dealing with different camshaft manufacturers. Increasing the duration raises the rpm at which the engine achieves maximum power. A stock early-production 302 Ford has an intake duration of 266 degrees and an exhaust duration of 244 degrees. This provides good low-end power, with the power peak being reached at less than 4000rpm. By increasing the duration to 320 degrees in the intake and 320 degrees on the exhaust, the peak power band is moved up to somewhere between 4800 and 8200rpm. When selecting a camshaft, depend on the specs provided by the camshaft manufacturer for the best duration for the peak power rpm band desired.

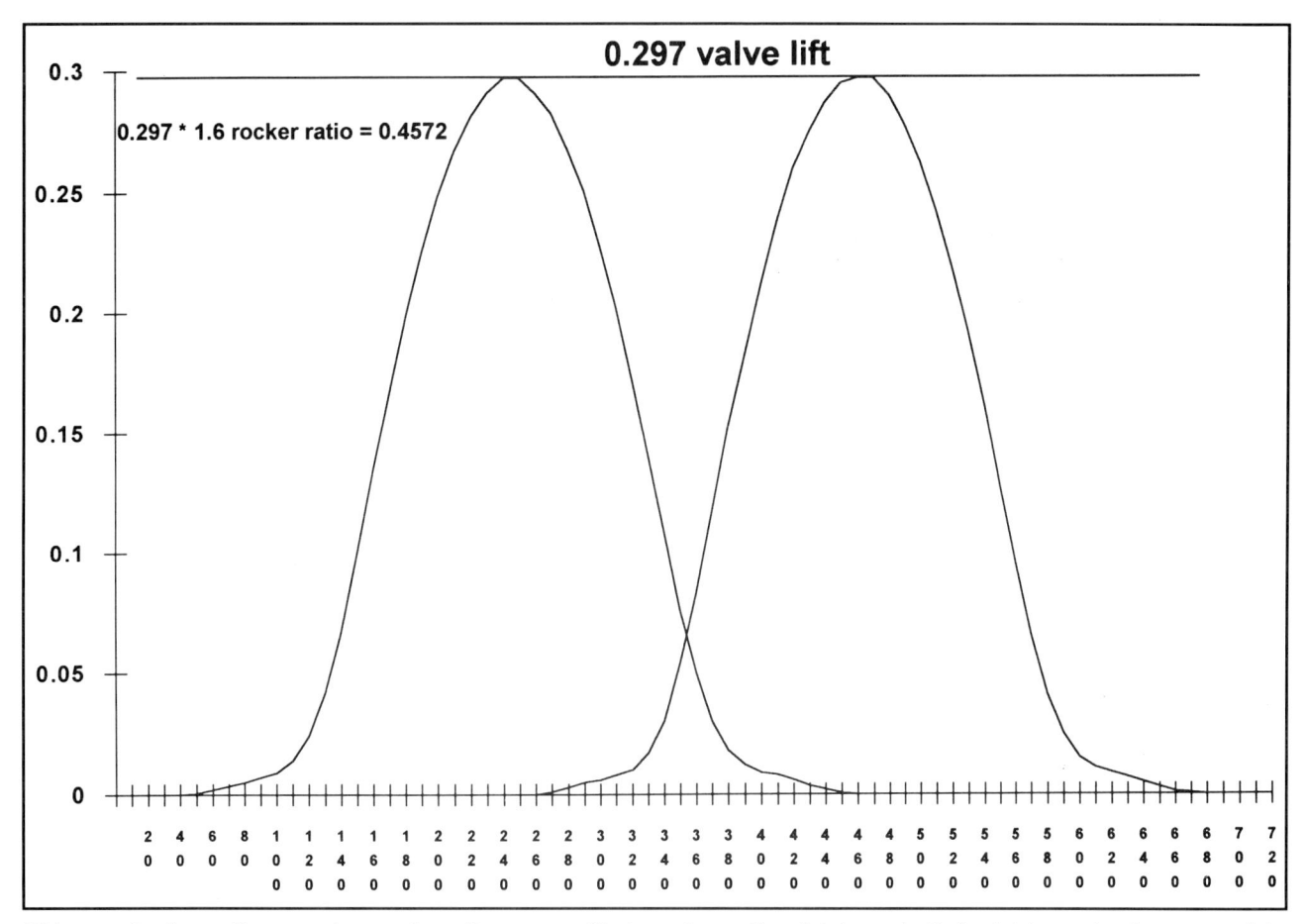

This graph shows the opening and closing of the valve on a cam with a lobe lift of 0.297in. The graph assumes the very unlikely rocker ratio of 1:1. The graph does show, however overlap and the fact that full operation of both the intake and exhaust cam lobe and valve requires 720 degrees of rotation.

Yes, unfortunately the performance industry and the Society of Automotive Engineers disagree about how duration measurements are to be taken. For the sake of the book, duration shall be discussed using the performance industry standards.

Valve Overlap

Valve overlap is the time at the beginning of the intake stroke when the exhaust valve is still open. Large valve overlap increases the efficiency of cylinder filling at high engine rpm. However, this increase in efficiency is achieved at the expense of reduced idle vacuum and therefore poor low-speed performance and a rougher idle. During the 1970s, increased valve overlap was used

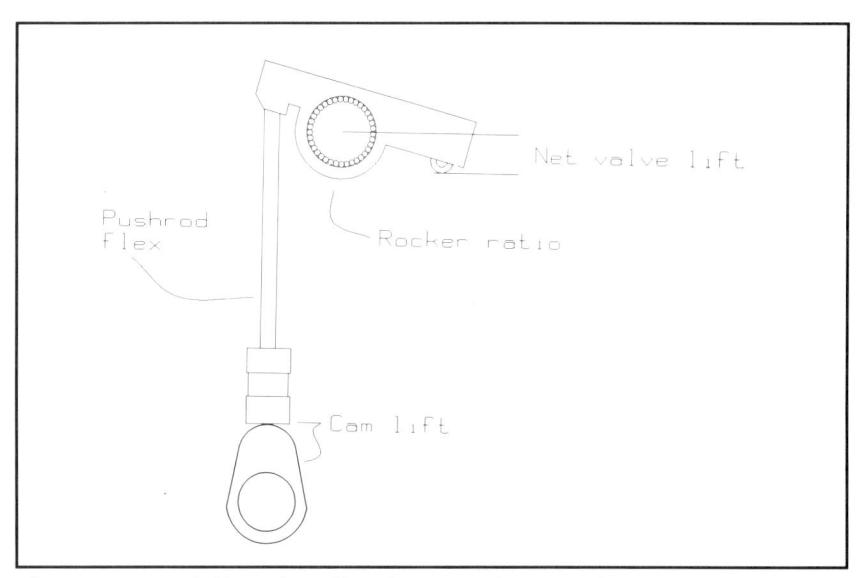

There are several things that affect the actual valve lift. The cam lift and the rocker ratio are the two major elements, and in an overhead cam application are the only factors. In a pushrod engine cam flex and pushrod flex can reduce the actual valve lift.

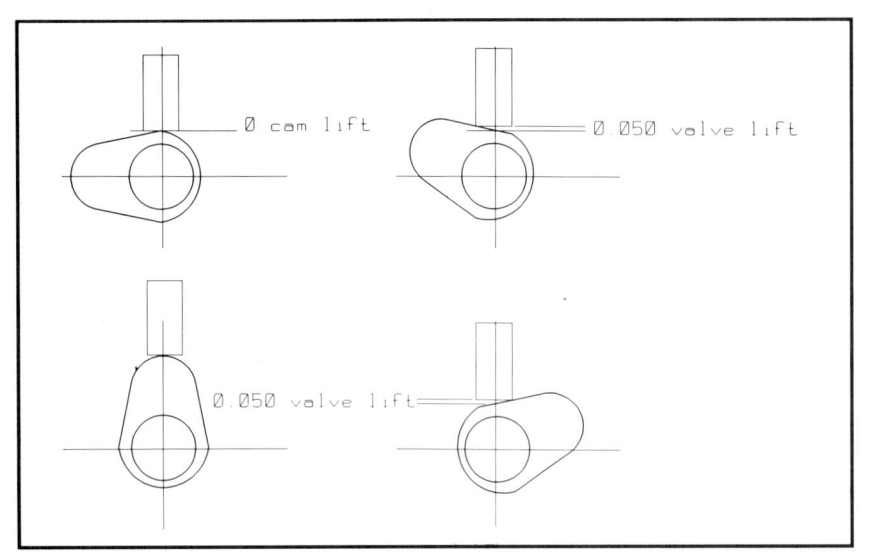

to draw exhaust gases back into the combustion chamber to cool the combustion process and reduce the emissions of oxides of nitrogen (NO_x).

In a performance engine, the greater valve overlap allows the average velocity of the gases flowing through the combustion chamber to remain high. The higher velocity increases the inertial flow of gases, ensuring that the cylinders fill better on each intake stroke and scavenge better on each exhaust stroke.

The negative side of increased valve overlap is the *lumpy* or loping idle which is characteristic of most racing engines. This lumpy idle results in erratic manifold vacuum, which

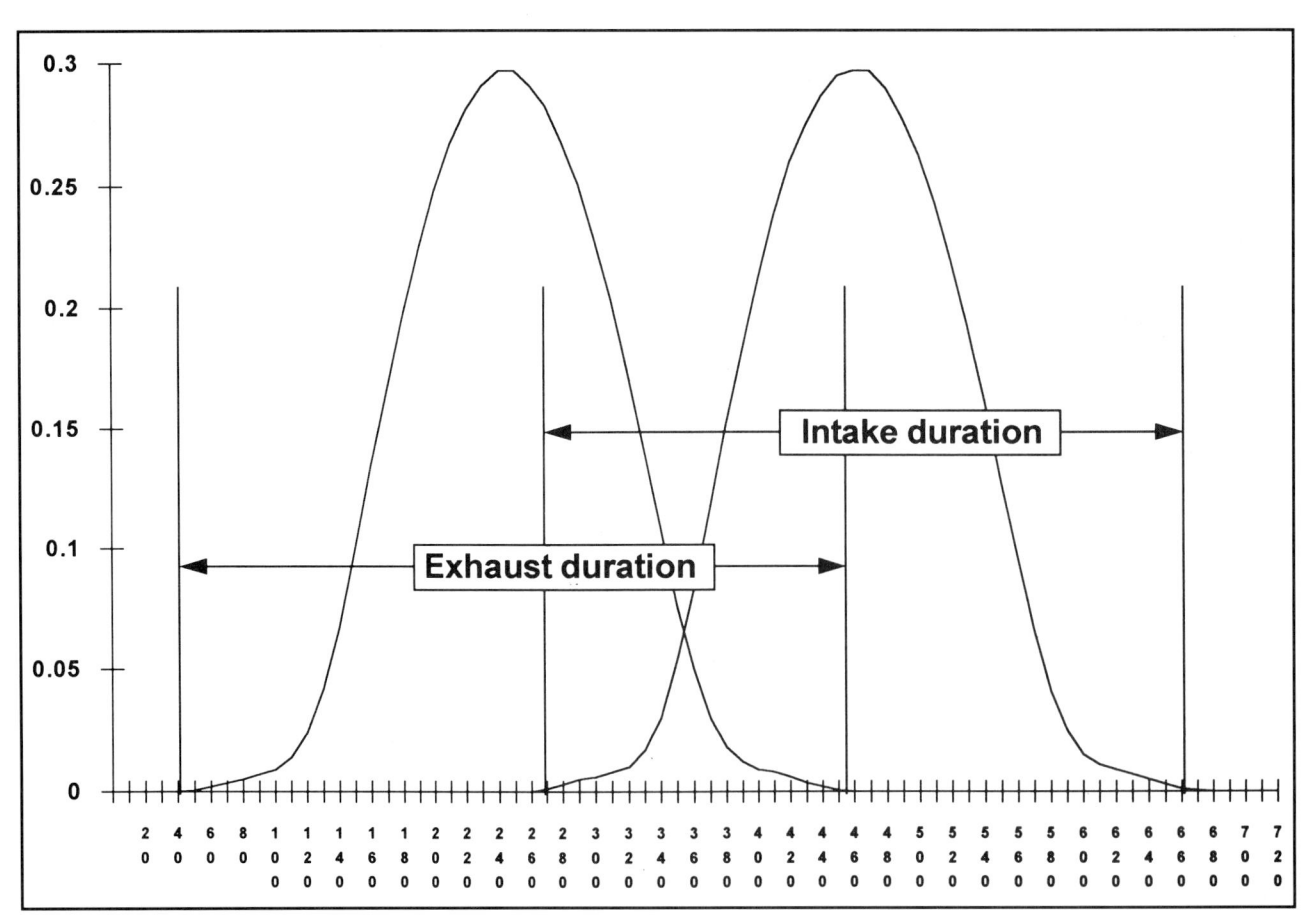

Duration is measured from an industry standard point of valve lift. That SAE standard for measuring duration is from 0.006in of valve lift. The performance industry uses 0.050in of valve lift.

Cam lobe duration is measured in degrees of crankshaft rotation. Remember that in a four-stroke engine the crankshaft rotates twice for every rotation of the camshaft. In the example above, in the area between 269 degrees and 455, both the intake valve and the exhaust valve are open. This is called valve overlap. Valve overlap aids in moving the air/fuel charge through the combustion chamber, the exiting exhaust gases aid in drawing in the entering air/fuel charge.

56

can seriously affect the operation of carburetors and fuel injection systems. Although NO_x emissions might be reduced at high engine speeds, this is at the expense of carbon monoxide and hydrocarbon emissions at lower speeds. Use of a large overlap camshaft on a street application will almost ensure that the vehicle will fail the local emission standards, not just in California, but in all fifty states.

A Less than Professional Example

Deep in the recesses of a garage in Fort Worth, Texas, I found a "racing" engine that had been assembled from junk parts in 1973. This racing engine started life as a 1967 Ford 289, was fitted with a crankshaft from a 1968 302, and a set of 12:1, 0.030in over TRW pistons. The camshaft that was installed had a 0.297in lift on both the intake and exhaust lobes. Since stock lifters had a 1.6:1 ratio, the valves could potentially open 0.4752in. And since solid lifters were installed, a potential 0.015in of valve lift are lost to the lash adjustment, and 0.003in to flex in the pushrod. This means that the net lift will be only 0.4572in.

The intake and exhaust ports of the cylinder heads must be able to flow a minimum of 83cfm per cylinder (55.58 x 1.5).

After browsing through the yellow pages for a shop with a flow bench, I located a small machine shop specializing in racing heads in Arlington, Texas, near my home. As the first cylinder head was fitted on the bench, the machinist noticed that the head must have come from the factory about 1962, mounted on a 221ci V-8 engine. At 5000rpm, this engine would only have required an airflow rate of 60cfm per cylinder (40cfm x 1.5).

$[(5,000 \times 221) / 3,456] \times 1.5 = 60$

If the airflow capacity of this head is no better than what was required for the 221, the 306ci engine will not receive sufficient air.

When the cylinder head was mounted on the flow bench, the airflow rate was set to 200cfm. Airflow readings were taken at progressive valve openings of 0.050in. The results of the test are listed in the table.

At zero inches of valve opening, the exhaust valve flowed at a rate of 13 percent of the 200cfm. The crude results of this test show an airflow rate of 26cfm. The manufacturer of this particular flow bench recommended a correction factor of 1.003 be used to adjust for inaccuracies in the bench. Thus, the adjusted airflow rate through the exhaust valve was 26.078cfm. (Although correcting for an error of 0.3 of 1 percent may seem nit-picky, nit-

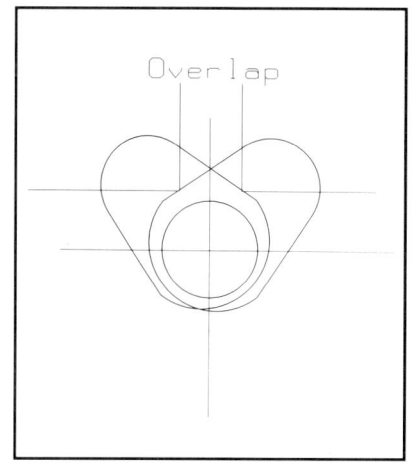

The arrangement of the intake and exhaust cam lobes allows the exhaust valve to remain open for a few degrees after the intake valve opens. This can not only improve intake air flow but during the 1970s was used to allow the dropping piston to draw some of the exhaust gases back into the cylinder. These gases cool the cylinder and reduce oxides of nitrogen emissions.

Valve Lift = (Cam Lift x Rocker Ratio) minus (Valve Lash + Rocker Flex)

The bore of this engine is 4.030in and the stroke is 3.0in, for a displacement of 306.13ci.

Displacement = P x Bore^2 x Stroke x No. of Cylinders
CFM = (RPM x Displacement) / 3,456

At 5000rpm, this engine will move about 443cfm of air.
This equals 55.38cfm of air per cylinder.

CFM per Cylinder = [(RPM x Displacement) / 3,456] / No. of Cylinders

---

Airflow Capacity

	Exhaust	Intake	Exhaust (cfm)	Intake (cfm)	20 Inches (cfm)
0.05	13.00%	13.00%	26.078	25.922	18.51538
0.1	20.00%	17.50%	40.12	34.895	28.4852
0.15	28.00%	37.00%	56.168	73.778	39.87928
0.2	35.00%	46.50%	70.21	92.721	49.8491
0.25	40.50%	57.50%	81.243	114.655	57.68253
0.3	43.50%	66.00%	87.261	131.604	61.95531
0.35	45.50%	69.50%	91.273	138.583	64.80383
0.4	46.50%	71.80%	93.279	143.1692	66.22809
0.45	48.00%	74.00%	96.288	147.556	68.36448
0.5	48.50%	76.40%	97.291	152.3416	69.07661

Deep in the recesses of a Fort Worth garage we found this engine. Built as a racing engine in early 1970s, it is your author's personal example of how to do everything wrong.

This is a rare device in many towns, and even in a metropolitan area as large as Dallas/Fort Worth they are few in number. However, a flow bench is necessary to confirm that any modification you have done to the head actually made an improvement.

picky is the difference between the front row of Indianapolis and spending the week before Memorial Day driving back to Houston.)

When the valve was opened to 0.10in, the flow rate increased to 20 percent. Using the above formula, the airflow rate became 40.12cfm. When the valve was opened to slightly less than the calculated exhaust valve opening height, the flow rate reached 96cfm. This is slightly better than the desired flow rate of 83cfm.

The problem with this reading relates to what goes on in the engine when the airflow increases toward wide-open throttle. As the throttle opening increases, the load on the engine increases and the engine vacuum drops. The readings in the fourth and fifth columns of the chart were taken at 24in of vacuum. The last column shows exhaust port readings at 20in of vacuum. There is a considerable difference here. As the throttle responds to the driver's demand for more power and more speed, the engine vacuum can drop to virtually zero. This may seriously inhibit airflow. The neat, scientific readings in the chart have little to do with real-world performance engines.

High-lift, long-duration camshafts and a wide-open throttle can team up to reduce the vacuum created by each cylinder. On a stock engine in good condition, each cylinder creates an average of about 20in of vacuum. The Allen Smart Engine Analyzer creates a vacuum waveform pattern. This pattern reflects changes in manifold vacuum as each intake

The first step of preparing the head to be flow tested is to remove the springs and valves. Install the valves in the ports to be tested. Install the wimpy little valve springs shown in the picture above to hold the valves closed during the test.

The head is then mounted on the flow bench. The flow bench creates a low pressure area (often and inaccurately referred to as a vacuum) below the cylinder head. The tool mount in the place of the rocker arm is marked off in thousandths of an inch and is used to incrementally open the valve. When the exhaust port is tested, as above, the flow bench blows the air through the exhaust valve port.

59

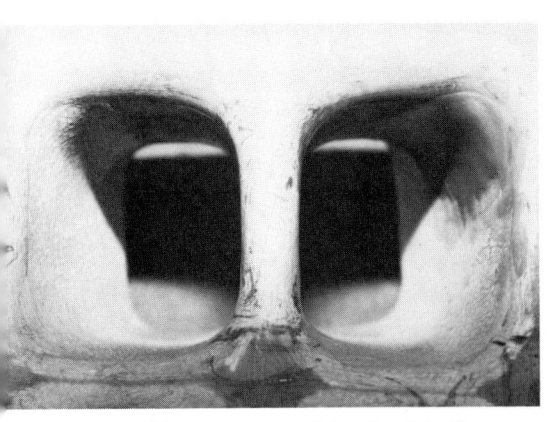

To most accurately simulate the presence of an intake and exhaust manifold modeling clay is used to smooth the flow of air in and out of the cylinder head ports.

valve opens and closes. When the intake valve opens, the piston has already begun its decent down the cylinder. This, in turn, creates a low-pressure area above the moving piston. When the valve first opens, the piston is beginning to accelerate rapidly so the low pressure (high vacuum) is at its greatest. As the valve approaches its maximum opening, the piston velocity begins to decrease and the low pressure begins to rise as the velocity decreases. If the throttle is wide open there is minimal restriction to airflow into the intake, and the intake fills with air rapidly and vacuum drops.

The point of all this is that there are limitations to the relevance of flow bench testing and other "ratings" of the air flow. At the very time the maximum air flow is desired, at wide open throttle, is the very time the pressure differential between the atmosphere and the cylinder is at its lowest. This means the vacuum is at its lowest. Because of this you always need to ensure there is a greater air flow capability than the engine seems to need mathematically.

In a gasoline-powered spark-ignition engine the ratio of fuel to air in the combustion chamber must be precise. Not only is the air used as an expansion medium to drive the piston downward, it is also used as a medium for evenly distributing the fuel particles in the combustion chamber. Igniting the fuel with the spark

The modeling clay is molded into a block of wood that has holes cut in it to correspond to the tubes and runners of the exhaust system and intake system.

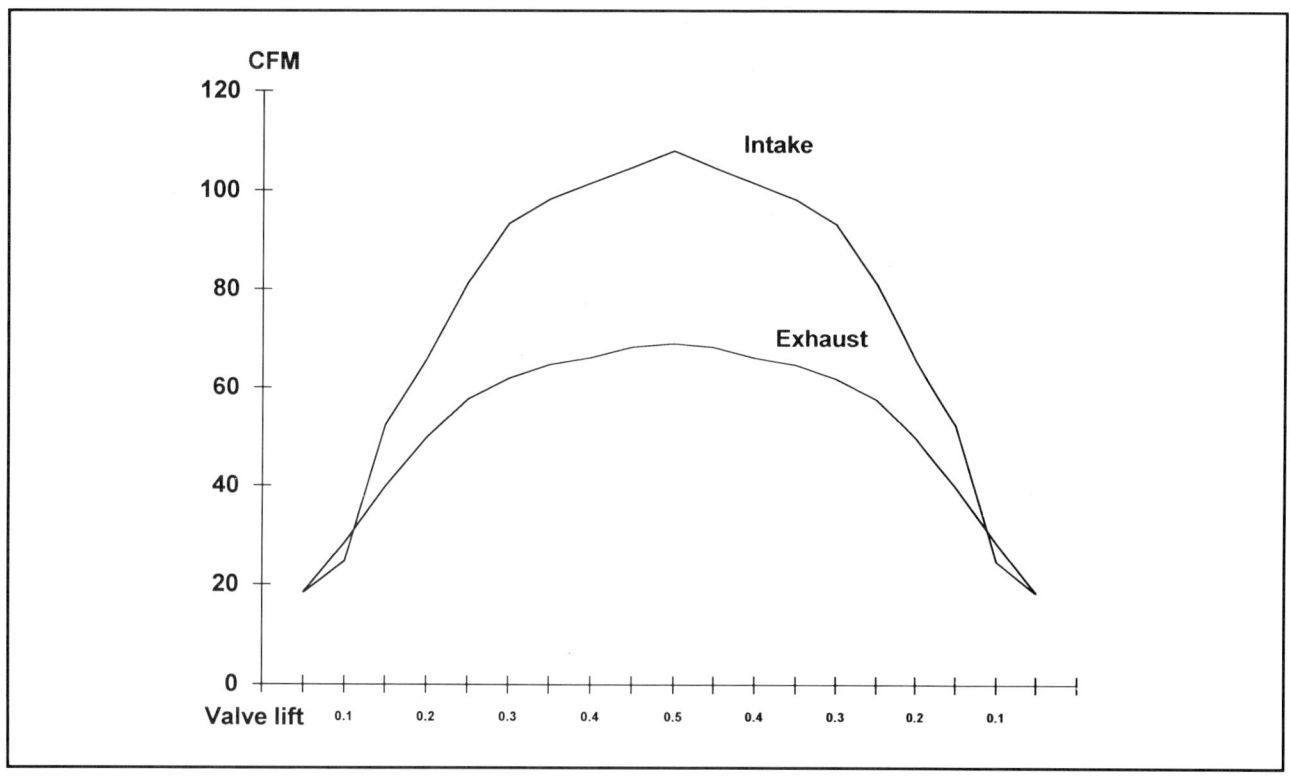

```
CFM
120
100                  Intake
 80
 60                 Exhaust
 40
 20
  0
Valve lift  0.1  0.2  0.3  0.4  0.5  0.4  0.3  0.2  0.1
```

Considering that the casting marks on this head indicated that it came from a 221, the flow through the head was surprisingly good. For these results we used a maximum flow rate of 200cfm and set the vacuum/pressure at 24in. The exhaust port would flow a maximum of 97cfm at 0.5in of valve lift. This engine could not create 24in of vacuum, especially not when maximum air flow is required. A more realistic figure of 20in of vacuum yields a maximum air flow through the exhaust port of 69in. This is the limit of what the engine under this head can push out. Since you cannot draw in more than you can push out, it is the limit of what can be pushed out.

plug only begins a process that must then be supported by the evenly distributed fuel itself. As the flame front travels through the combustion chamber it leaps from one fuel droplet to another, much like a forest fire leaps from one tree to another. If there is a gap in the trees, a fire-break will stop at the break. If there is a break in the fuel particles, the flame front will stop at the break. An inadequate amount of fuel will result in large gaps in the fuel particles. These large gaps will then result in some of the fuel not being burned. Unburned fuel results in high emissions and a loss of potential power.

Notice the blobs of clay where the Plexiglass tube joins the Plexiglass panel. These blobs were placed there by the flow bench operator to seal tiny holes in the junction between the tube and the panel. Any air leaking into the head between the wood/clay "manifold" and the base of the flow bench can destroy the accuracy of the test.

Warped and Cracked Heads

8

In an ideal world, everyone would contribute to charities, pizza would be low in fat, and cylinder heads would never warp or crack. Those with the luxury of bank accounts the size of Ross Perot's are not concerned about a warped or cracked cylinder head for their performance engine. These people simply buy a new, nonwarped head. But for the rest of us, there are some things to consider.

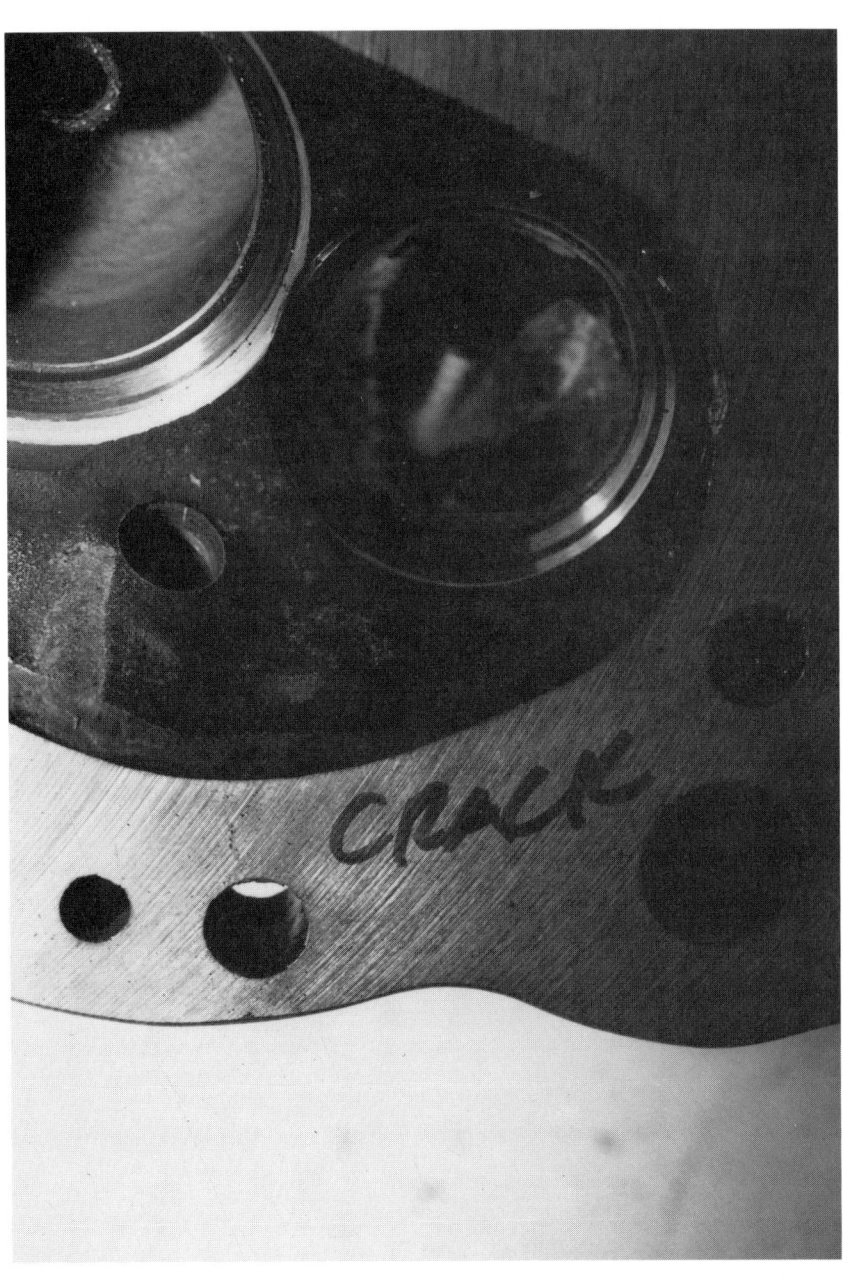

Repairing Used Heads

The likelihood of buying a used head that is free from damage is directly proportional to the rarity of the head or engine.

If the engine is a Chevy 350, the easiest and best way to repair a damaged head is to replace it. Some of us, me included, lack the luck or wisdom to work only with small-block Chevys. We find ourselves taking on projects like a pro-stock Skoda (a Czechoslovakian auto). You do not find Skoda heads in every wrecking yard in America. The fact is, cylinder heads do get damaged, and may be too rare or expensive to replace—they must be repaired.

Before purchasing a used cylinder head, carefully inspect it for any evidence of cracking or damage. Begin with the assumption that the head is cracked. If you cannot see any cracks, then you can be sure the cracks are hidden. In order to prove the head is not cracked, have it pressure tested.

White smoke can indicate a problem with the head. If the engine that the cylinder head came from was producing white smoke before it was torn down, be cautious. White smoke has two common sources, transmission fluid and antifreeze. If the car is equipped with an automatic transmission, look for a vacuum

Can you see the crack in this cylinder head? It runs from the headbolt hole to the left of the "C" in "CRACK" toward the combustion chamber. Cracks like this are hard to spot and are specifically designed by the great powers of the ether to only cause a problem halfway across Raton Pass between New Mexico and Colorado... or just before the cash run at the Friday night drags.

modulator. You will find the vacuum modulator located toward the back of the transmission, next to the output shaft housing. It is easy to recognize because there is a vacuum hose going into it. Remove the vacuum hose and inspect for transmission fluid in the hose. If there is transmission fluid in the hose, replace the vacuum modulator. This is the most likely source for the white smoke.

If there is no transmission fluid in the vacuum hose or if the car is not equipped with an automatic transmission or does not have a vacuum modulator, the most likely source of the white smoke is a blown head gasket, cracked cylinder head, or cracked head. This can be confirmed with a compression test, or better yet, a cylinder leakage test.

Cylinder Leakage Testing

The cylinder leakage test is superior to the compression test, although its goals are the same. It also requires more equipment. With the piston at top dead center (TDC) compression stroke, insert the hose of the cylinder leakage tester into the spark plug hole of cylinder number one. Do not connect this hose to the pressure tester yet. Connect the other hose, the one that will not fit into the spark plug hole, to a compressed air source. Adjust the knob of the compression tester to zero percent, and connect the hose in the spark plug hole to the tester. Measure the percentage of leakage. Use a piece of heater hose held to the ear to determine where the leakage is. Then stick the heater hose into the open bore of the fuel-injection throttle assembly or carburetor. If there is a great deal of air escaping through there, you have an intake valve problem. Stick the hose in the exhaust. If there is excessive air escaping, you have a bad exhaust valve. Next, observe the radiator coolant. If there are a lot of bubbles in the

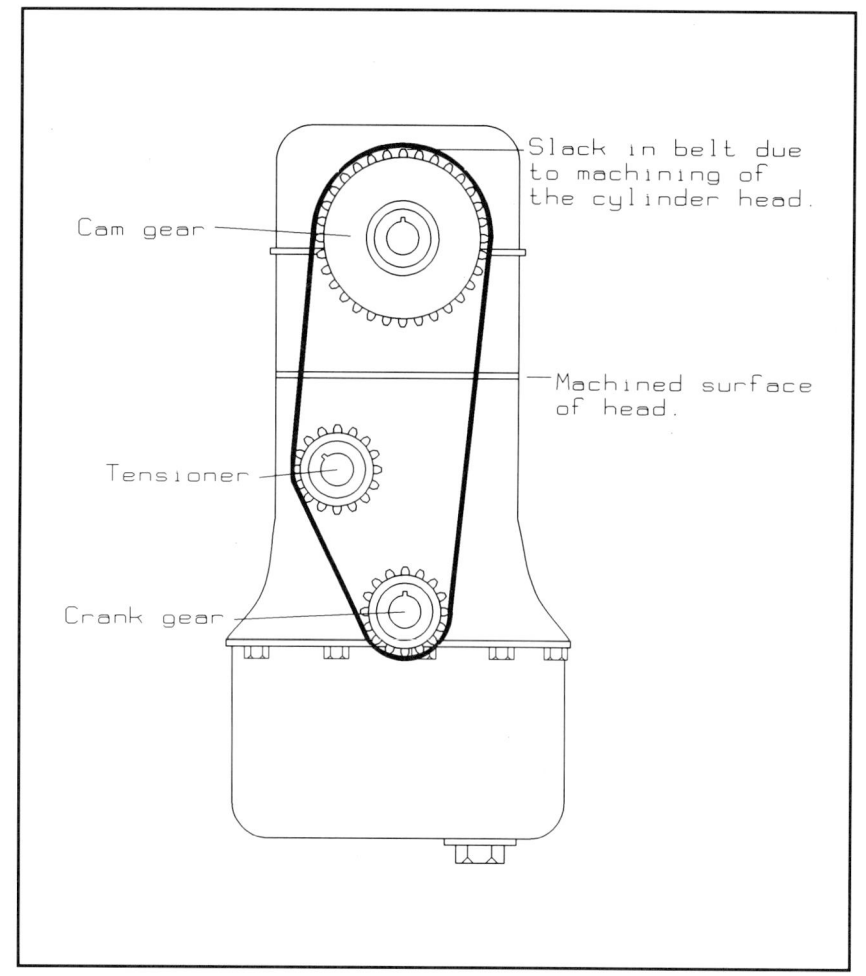

As a fellow who cut his teeth on Detroit iron in the 1960s and early '70s I always thought that you stopped removing metal from the bottom of the head only when the combustion chamber disappeared or when you could see water jackets. I was totally unprepared for the first Datsun 240 engine I did performance modifications on. I had the head machined 0.030in. The machinist I had do the work smiled knowingly as he handed me the head across the front counter. After installing the head, the cam gear, the crank gear, and the timing chain, I discovered to my dismay that the chain was looser than Richard Nixon's jowls. The compression ratio, however, was exactly what I had planned. The next morning Al, the machinist, called and asked if I would like to buy some 0.030in cam tower shims.

coolant, then you have a blown head gasket, or a cracked head or block.

It is relatively easy to determine whether the white smoke is burned transmission fluid or antifreeze. The antifreeze has a distinct, *sweet* odor while the transmission fluid just smells smoky.

Pressure Testing

Perhaps the most important operation that can be done by the machine shop is testing the integrity of the water jacket in the head and head. Most machine shops are equipped to do pressure testing.

To begin the pressure test, all of the water ports in the head or head are plugged. Air or water is forced into the cylinder head or head If air is being used, soapy water is sprayed on the surfaces in question. If there is a leak, air bubbles will be seen.

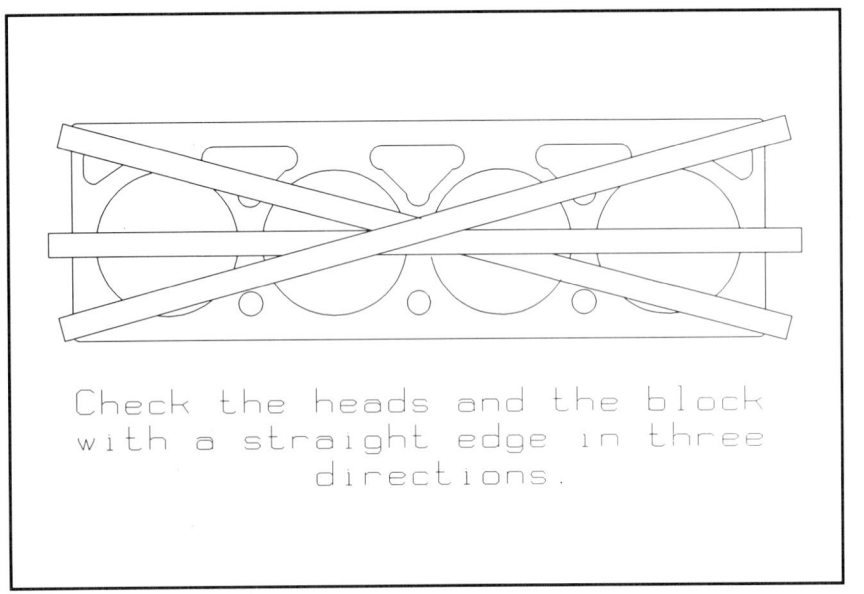

Check the heads and the block with a straight edge in three directions.

Many years ago I announced to my girlfriend (later to be my wife) that I wanted to be a race car driver. She asked if I had a warped mind. Well I do not know about that, however, it must have been a curse because since then I have been plagued by warped heads. When you purchase a used head or consider the use of an old head, check the flatness of the head carefully with a straight edge. The yard stick from the village hardware store will not do the trick, so buy an appropriate straight edge from one of those guys on a tool truck.

Aluminum heads sometimes become porous. When this occurs coolant will literally seep through the metal when pressurized. Should your machinist find this situation he can sometimes seal it with a resin. Like so many other things in rebuilding an engine, because of the amount of work required to repair mistakes or bad judgment, a porous head should just be replaced.

Magnetic Crack Inspection

This technique works only on ferrous metals as it depends on the dispersion of the magnetic field through the metal. The principle is similar to that famous high school physics experiment we discussed earlier, the one that illustrates magnetic flux fields. In that experiment, a magnet is placed under a sheet of paper and iron filings are sprinkled over it. As the iron filings fall, they line up along the magnetic fields of flux. The same basic technique is used In magnetic crack inspection. Iron filings are sprinkled on the head. A large electromagnet is then held a fixed distance above the head. Cracks in the cylinder head create distortions in the magnetic field, which affects the alignment of the iron filings.

Dye Crack Testing

Remember the old, cracked coffee cups at the roadside diner we discussed earlier? Remember, we said that if you examine one of the cracks in the cup, you will find there is residue from the coffee of customers who sipped a little java while waiting for the bus to take them off to World War II.

This illustrates the principle of dye crack testing. A penetrating dye is sprayed on the surface of the item being tested. After being allowed to soak in and dry, a cleaner is used to remove dye from the surface. Like the coffee stains in the crack of the cup, the dye will remain in the cracks of the component being tested. The dye method is especially effective on nonferrous metals such as aluminum and magnesium.

Fixing a Cracked Head

Cracked heads are common in some applications, climates, and uses. They are rare in others. If the head is cracked, there are several repair methods available to you, welding and threaded plugs among them.

Welding the head is a highly specialized art; most machinists recommend simply replacing the head. Welding is especially difficult on aluminum and alloy heads. If the head is a rare application, consult your machinist for recommendations on who is qualified locally to do the welding.

Sealing the crack with threaded plugs works well when the crack is small and replacement or welding is out of the question. In this technique the machinist will stop-drill the crack, then drill a series of large holes along the length of the crack. Special threaded plugs are then threaded into the holes. When threaded as far as they will go, they are cut off and ground flush with the head surface. Consult your machinist for advice on repairing cracked heads.

Resurfacing Warped Heads

Constant changes of temperature, coupled with inadequate maintenance such as routine retorquing of heads, or incorrect retorquing of heads, can result in warping. Of all the machining operations performed on a cylinder head, resurfacing may be the most common.

Resurfacing the head involves milling off several thousandths of an inch of metal. In effect this removes the high spots, making the head flat again.

Several factors must be taken into account when the head is resurfaced. If the engine is an overhead cam design, removing several thousands of an inch of material will cause the camshaft or camshafts to be closer in proximity to the crankshaft. This can affect the tension or amount of slack in the timing chain or belt. Additionally, a warped cylinder head bottom may indicate that the top of the head is warped as well. If the top is warped, the cam bearing may be out of alignment. Misaligned cam bearings can bind, damage, and even break the camshaft.

Solutions? First of all, you need to realize that taking a link out of the chain to shorten it is not the answer. Within one or two revolutions of the crankshaft, the camshaft would be far enough out of phase to bend every valve in the head. And don't doubt that some people actually attempt to solve the problem by literally removing a link. I once saw a professional technician in a hurry to finish a job do that very thing. He knew better, he just was not thinking. So, think!

For many applications, automatic chain and belt adjusters will take up a great deal of this slack. When too much metal has to be removed, the slack can be corrected by shimming the cam bearing towers if the application you are working on has removable cam towers. These shims are available from your machinist or, in some cases, from an appropriate dealer. For the applications where the cam towers are not removable, your machinist might be able to line bore the tower cam journals and install oversize cam bearings. There are head designs that do not lend themselves to this solution, either, and will simply have to be replaced.

In the late 1970s, I served a brief sentence for a Mercedes-Benz dealership. When it came to a number of the diesel engine applications, one could almost confirm the date on the calendar by the line of cars waiting at the door in early spring to have their cylinder head gaskets replaced.

Please understand that this high incidence of head gasket replacement is inherent to the Mercedes. Rather, it was largely a result of lack of proper maintenance and the extreme high temperature conditions in north-central Texas.

As this season approached in 1979, a local hanger-on dropped in. (For those who do not know, a hanger-on is an accountant or doctor who wanted to be a mechanic, but whose parents forced him to be a *professional*.) Anyway, this particular fellow was a Mercedes owner whose head (cylinder, that is) had been replaced the previous year. This was particularly memorable for everyone in the shop because of how loud his wallet screamed when it opened. The guy was trying to sell us a service. He proposed to straighten cylinder heads by baking them in an oven. As we listened, the muffled snickers were noticeable to all. As much to get rid of him as anything, we gave him one of the cylinder heads that had been condemned for replacement. It worked! Whether he invented the process or simply heard about the idea before we had, the practice is now commonplace.

Straightening the head may be a viable alternative that eliminates many of the problems already addressed. However, the technology is not available everywhere, and the success ratio seems to vary a great deal with the skill and knowledge of the technician.

The process involves slowly heating the head to a high temperature (with the head bolted to a machined stainless steel flat plate). While this may seem simple enough to do in your oven at home, overheating the head can soften the metal or make it brittle, while not applying enough heat will prove ineffective. Also, if the head cools too quickly the consistency of the metal can change, or it may warp and re-

lease like a spring when the machined plate is removed.

Misaligned cam bearings on overhead cam engines can result from either the original head warpage or from the efforts to repair the warpage. This is especially true of the baking method. Before the cylinder head is installed, be sure to fit the camshaft into the head to ensure it turns smoothly. If you send the cam along with the head, the machinist correcting the warpage will check this for you as well.

To properly check the overhead cam head for misaligned cam bearings, it must be bolted to a torquing plate or to the block. This is especially true if the head's mating surface with the block is not to be resurfaced. Your machinist should be able to line bore the cam journals of the towers off-center, and install oversize cam bearings.

After all this effort to ensure the cam turns freely in the head, there is still one variable. Imagine a valve job on a 1978 Honda Civic CVCC. The customer's bill is already a microcosm of the national debt. Most of this is due to the customer's own desire for attention to detail. In any event, on this particular car the head-to-block surface of the cylinder head had been machined. The free movement of the cam had been checked with the head bolted to a torquing plate—perhaps the only CVCC-compatible torquing plate in the entire Seattle area at the time. The head was then bolted onto the block, and the cam and timing belt were installed.

As I began to adjust the valves, the crankshaft would not turn. Several horrible thoughts crossed my mind, as well as several unprintable expletives about the condition of the crankshaft. As it turned out, the problem was worse than I had first imagined. The block and crank were fine, at least on the lower end. However, the block's mating surface with the head was se-

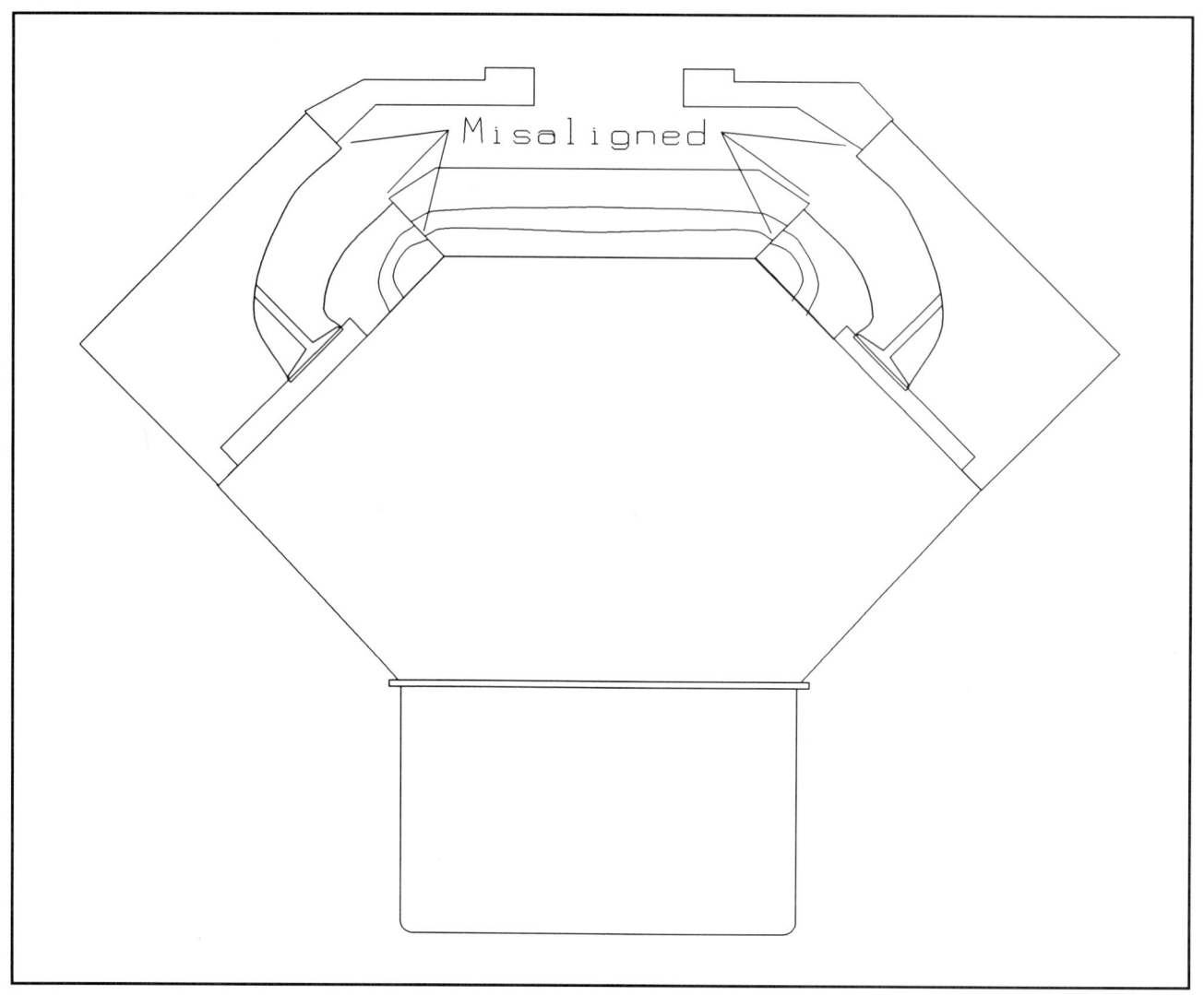

When the heads are milled on a "V", they are lowered with respect to the intake manifold. This can cause a misalignment of not only the intake ports as mentioned earlier but also of the coolant and in some cases oil passages.

verely warped. When the head was bolted on, it had conformed to the shape of the block with absolutely no argument. The block had to be replaced, which was cheaper than the alternative of being resurfaced, line bored, and the deck height reset. Instead of a profitable complete engine overhaul, I ended up billing for only part of the engine rebuild and had to eat the price of the replacement head. So take warning: Check the surface of the block as soon as the head is removed.

Checking for Block Deck Warpage

With a flat file, clean the burrs and old gasket material off the block deck surface. Place the straightedge on the block's mating surface with the cylinder head. Using a feeler gauge, check this surface for warpage. Three measurements are advisable: two at an angle across the deck, and one parallel to the longitudinal centerline of the block. It is quite uncommon for the block to be warped; cylinder heads warp far more often. The clearance measured with the feeler gauge should be less than 0.002in.

If the engine is a V-design, both cylinder heads should be machined the same amount. Since resurfacing the cylinder head alters the size of the combustion chamber and affects the compression ratio, machining the heads differently will create unequal power and performance characteristics from each side of the engine. Additionally, removing metal from the heads can affect the interference angle be-

66

tween the intake manifold and its mating surface on the cylinder head. No more than 0.024in should be removed from these heads. If the head is badly warped, the mating surface of the cylinder head will need to be machined with the intake manifold. Not doing so creates several potential problems:

1. The ports of the intake manifold may not line up properly, which can cause air leaks (often, and incorrectly referred to as vacuum leaks), oil leaks, and coolant leaks.

2. Since the rocker arm shaft or studs are closer to the camshaft, the rockers may bottom out the valve springs. Note that this problem is *not* unique to the V-4, V-6, and V-8; it can happen on any overhead valve design pushrod engine.

3. Removing metal from the mating surface of the head and block increases the compression ratio. Increasing the compression ratio increases the potential for detonation or pinging.

Imagine yourself after having spent $1,000 or more in the pursuit of excellence, then discovering that the boss can hear you accelerate up the hill more than five blocks away when you are late for work. The oil companies are shifting their emphasis away from anti-knock additives and toward emission control additives. Make every effort to retain the same compression ratio—unless you are willing to take some extraordinary steps to prevent spark knock. Raising the compression ratio may require the use of the most expensive gasoline available, or the habitual use of a spark prevention additive.

The simplest solution to this problem is to use a specially designed shim or a thicker head gasket. Fel-Pro Gaskets, for one, makes a 0.020in shim for several different engines. Fel-Pro and other gasket companies also make head gaskets in various thicknesses. Check into these options at your local parts house, or consult your machinist. This solution solves *all* of the potential problems.

Here is another option: If shims or thicker gaskets are not an alternative for the application you are working on, the intake manifold can be machined to ensure a perfect fit. This machining still leaves the problem of the shorter pushrods being required. Refer to the chart in chapter 6 for how much to machine. Based on the head angle, and the amount machined from the head, the chart tells you how to determine how much to mill from the manifold-to-head-surface (For specific degree of head angle, multiply [amount milled] x N = amount to mill from manifold-to-head surface.)

Doing so still leaves the problem of the shorter pushrods being required.

For people born with a Hurst Shifter "T-handle" in their right hand, and who are impressed with flame-breathing monster iron, they do make shorter pushrods.

Cylinder Head Measurement

9

Removal and Measurement of the Cylinder Head

One of the most important parts of disassembling a head is to do so with your eyes open. As a shop manager and owner I found this approach to be missing even from the best of professional technicians. Carefully inspect each component as it is removed from the head. Disassembly is the time to clean and measure parts.

(This section describes a typical teardown procedure. It is intended to be a guide, not a precise description. The exact sequence of disassembly and procedure you use on the head you are rebuilding will vary with head design.)

Begin the disassembly by removing the valve cover. Inspect the valve cover for cracks, damage, and warpage. If necessary, start a list of additional parts with a new valve cover. Remove the rocker arms or rocker arm assembly next.

Inspect the rocker arms. On overhead valve heads there are three common rocker arm designs.

Rocker Design 1

The first design is a stamped steel rocker arm that mounts on studs protruding vertically from the cylinder head. The rocker arm is held in place by a nut. To facilitate rocking, this nut has a rounded bottom.

When removing these rocker arms, check them for cracks in the area where the ball nut makes contact. In spite of their low-cost design, this rocker arm has few failures. The most common failure is when one of the nuts allows the tension created by the valve spring and the pushrod to force its way through the floor of the rocker arm. The second most common failure is when the pushrod is forced through the end of the rocker arm. Inspect this mating surface equally well. Wear can also occur on the machined surface when the rocker makes contact with the end of the valve stem.

Even a pro occasionally misses a deteriorated rocker arm. If there is any doubt in your mind about the condition of the rocker arms, add them to your list of things to buy.

Rocker Design 2

The second rocker arm has a shaft that runs from one end of the cylinder head to the other. This design features cast rockers and an oiled bearing surface between each rocker arm and the shaft. The rocking surface of this type of rocker arm is pressure-lubricated and therefore tends to be more dependable than the ball nut-mounted arm.

During disassembly, inspect both the bearing surface of the rocker arm and the area of the shaft where that bearing surface mates. If either the shaft or any of the rocker bearing surfaces show any signs of wear, replace the shaft and all the rockers. Also inspect the surface where the rocker contacts the top of the valve stem. If the shaft is in good condition, replace only the defective rocker. If more than one-third of the rockers show significant wear on any of the wear surfaces, replace all rockers and the shaft.

Rocker Design 3

The third rocker arm design is much like the second. A single

The most important part of rebuilding the cylinder head is disassembly. This should be done with your eyes open. The old head gasket, for instance, can give you good clues concerning the condition of the head and block. Breaks in the metal portions of the gasket surrounding the cylinders might indicate grooving, cracks, or other damage to the head.

shaft supports and provides the pivot point for all of the rockers. Unlike the second design, however, a rocker of this type has no actual bearing. This stamped steel rocker has a machined area which pivots on the shaft.

Check the surface which mates with the shaft for evidence of wear. Inspect the shaft. Make sure the pushrod cup is in good condition as well as the rocker arm surface that comes in contact with the valve stem. If the shaft shows evidence of wear, replace the shaft and all the rockers. If the shaft is in good condition, replace only the worn rockers.

Pushrods and Lifters

For each cylinder, punch two holes in a piece of cardboard. Re-

If you plan on reusing the original rocker arms inspect them carefully for wear and cracks. If you plan on using the engine in a racing or performance environment, then roller rockers should be strongly considered.

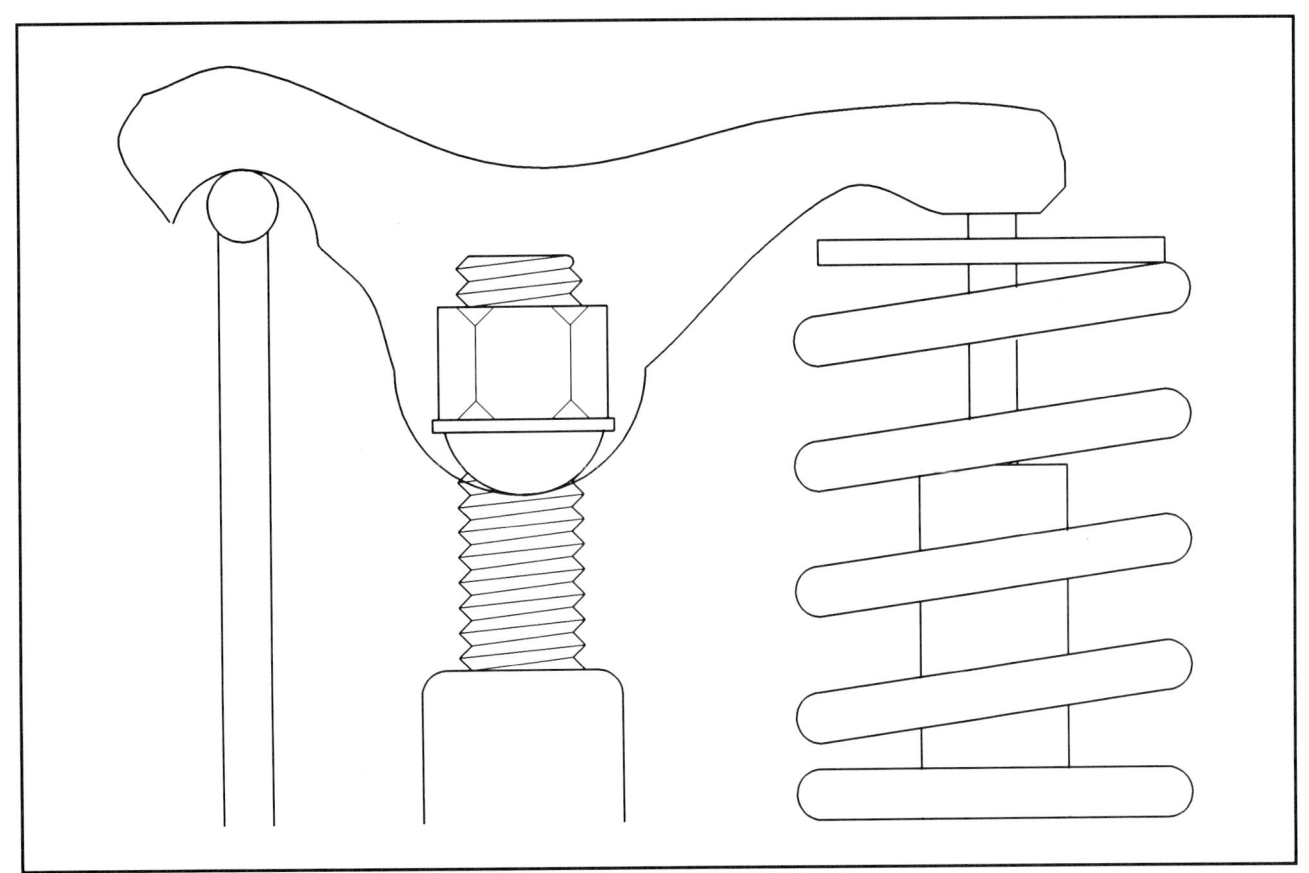

One of the more common rocker arm designs for overhead valve engines is the ball/stud rocker. The point where the rocker pivots on the ball is under extreme stress. For stock applications this design works extremely well and has a long life expectancy. But add a high-lift, long-duration camshaft with high-tension valve springs and rpms approaching 5000 sustained, and this type of rocker will not do the job.

69

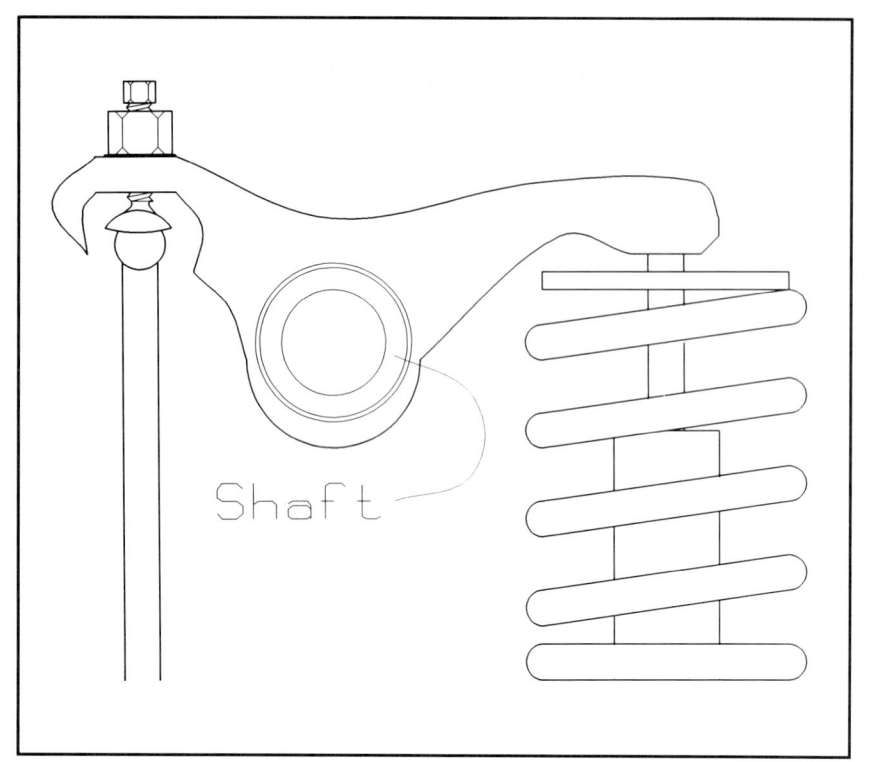

This cast iron, shaft-mounted rocker incorporates a bearing surface to reduce wear and increase durability.

This rocker will perform well for most any application short of the racetrack.

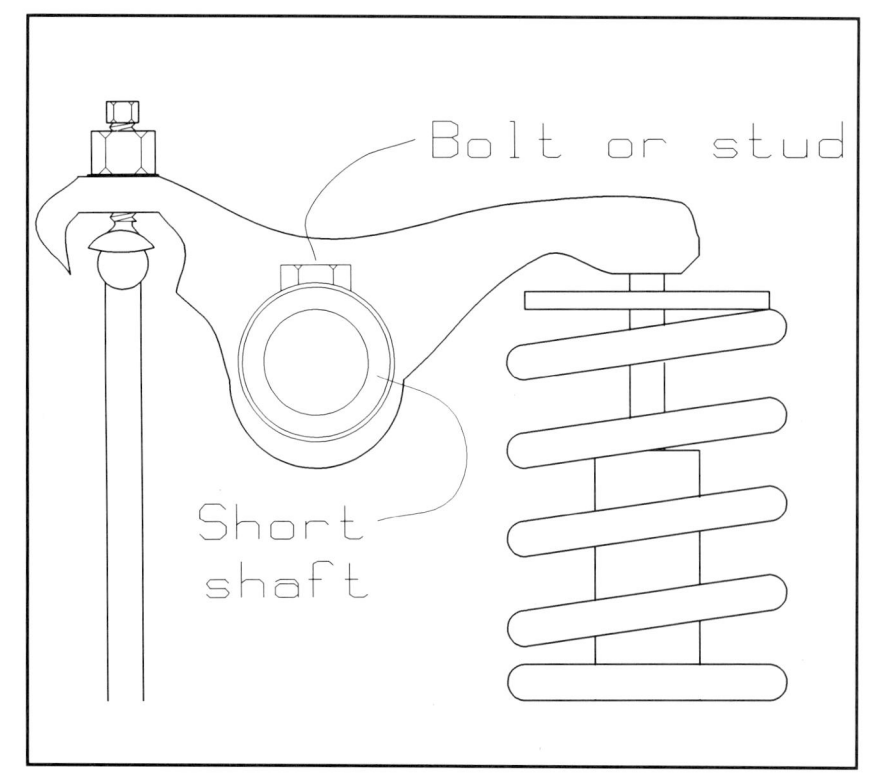

Similar to the shaft-mounted rocker, this rocker is also mounted on a shaft.

This shaft is very short, however and will support only one or two rockers.

move the pushrods from the engine and push them through the cardboard, taking note which cylinder each is from and whether it is the exhaust or intake valve. After all the pushrods are removed, lay a clean piece of glass on a flat surface. Roll the pushrods one at a time across the glass. Any that do not roll easily are bent and should be replaced.

Lifters

If the engine is a V-6 or V-8, remove the intake manifold. Remember that if the cylinder heads on these engines are resurfaced, it may be necessary to machine the intake manifold, too. Failure to machine the manifold mating surface to the cylinder head may result in vacuum leak.

Remove the lifters from the block. If the lifters are to be reused, they must be kept in order. A small box, just big enough for the lifters to fit in two rows for V-6 and V-8 applications (or a single row for inline applications) will work perfectly. Label the box left-right, front-back. During an overhaul, do not replace the camshaft without replacing the lifters and do not replace the lifters without replacing the camshaft.

Overhead Cam

Overhead cam heads come in two common designs with respect to the actuating of the valves. In the first design the camshaft sits to the side and applies pressure to rocker arms that open and close the valves. In the second design the camshaft sits immediately above the valves and opens and closes them by way of a cup and shim.

Rocker Arm Style Overhead Cam

On many overhead cam engines it will be necessary to remove the timing cover and remove the timing chain before removing the camshaft. Be sure and draw a diagram of the alignment of the timing chain and gears before removing the chain. The rocker arm overhead cam design places the camshaft to the side of the valves. A rocker arm is supported by a ball-tipped stud on one end and by the end of the valve stem on the other. The cam contacts a flat piece of the rocker arm. The downward pressure of the cam against the rocker pivots it on the ball stud, forcing the valve open.

When removing, inspect the machined surface on which the cam rides for evidence of wear and pitting. In many engine designs, these machined surfaces are lubricated very poorly. Also check the ball cup surface that mates to the ball stud. Wear on the machined surface that mates to the valve stem is also common. It is a fairly safe bet that you will find enough damage to warrant replacement of these rockers. Take this opportunity to carefully inspect the camshaft lobes for wear, rust, pitting, and other damage.

Shim and Cup Style Overhead Cam

Another overhead cam configuration is the shim and cup. A cup fits over the end of the valve which is recessed in the head. An adjustment shim fits in the top of the cup. As the cam rotates, it presses on the adjustment shim, which forces the cup downward, opening the valve. Remove the camshaft and inspect. remove each to the valve shims. In most situations, these shims will be reusable. Inspect and put them aside until reassembly. Please note that the chances of the same shim fitting in the same place during reassembly are pretty slim.

Inspect the sides of the cup for damage. Check for damage where the cup slides into the head. Make sure that the cup moves smoothly in the head. If it does not, be sure to tell the machinist when you have the valve work done on the head.

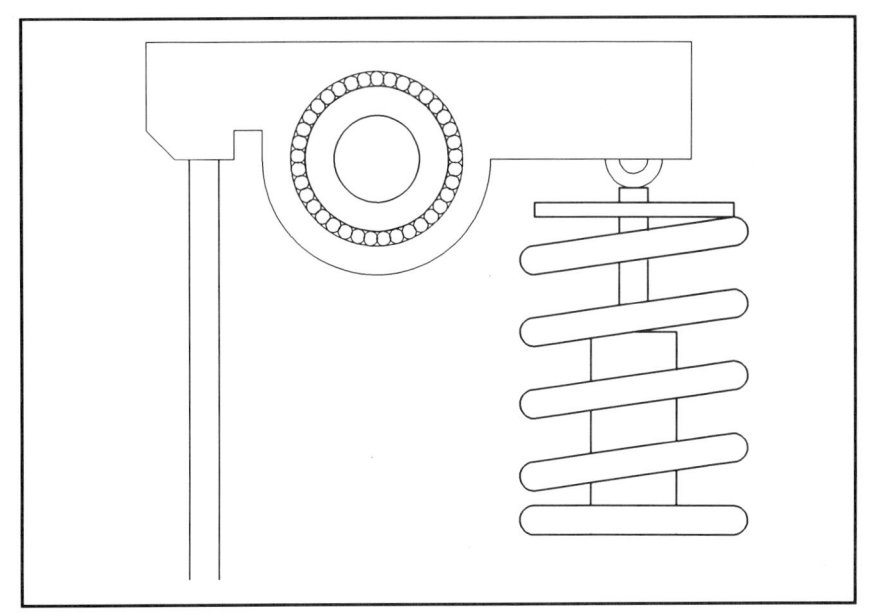

The roller rocker is the ideal replacement for any rocker. On grandpa grocery store car it means that the upper portion of the valve train might outlive junior. On the performance application it provides durability when used in conjunction with high-tension valve springs and high-lift camshafts.

Removing Valves

Although, as previously implied, there are ways of removing the valves that do not require a valve spring compressor, safety demands that I advise the use of one. Compress the valve spring with the compressor, then carefully remove the valve keepers and put them in a safe place. It is a little known fact of head rebuilding that an extraterrestrial

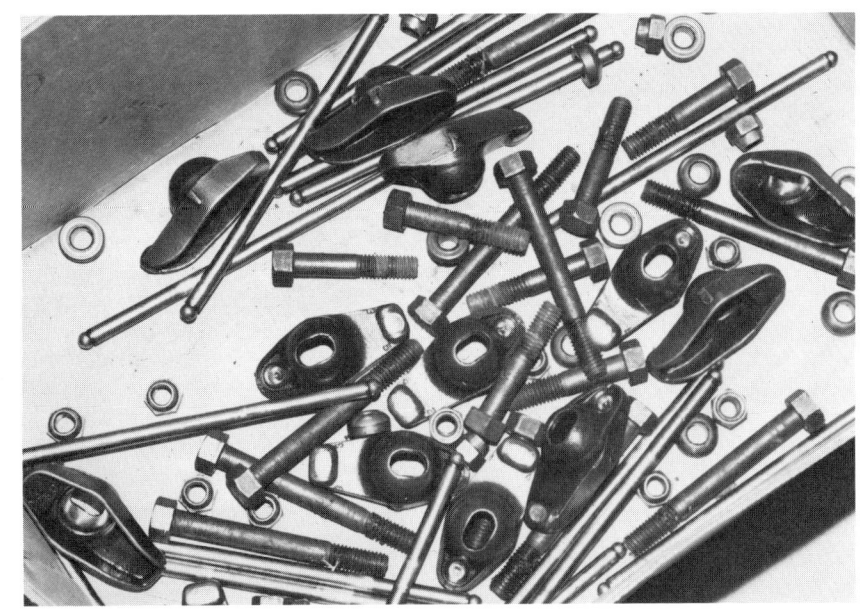

In this box lies the refuse of a head overhaul. The rocker arms have been replaced with roller rockers. The pushrods have been replaced with shorter pushrods to compensate for the milling that was done to the heads and the head bolts should always be replaced.

If the original rockers are to be reused check carefully for signs of wear and cracking. If the rockers are of the pivot ball type, check especially for wear in the area where the pivot ball contacts the rockers. This one was nice and shiny at the contact point, but not cracked or excessively worn.

Often stress cracks can be seen on the underside of the pivot ball rocker. This one appears to be free of cracks. Note, however, the extreme wear in the area where the rocker contacts the valve tip. This rocker was replaced when the camshaft was replaced and the valve spring were replaced with high-tension springs. It literally has less than 40 hours of operation on it.

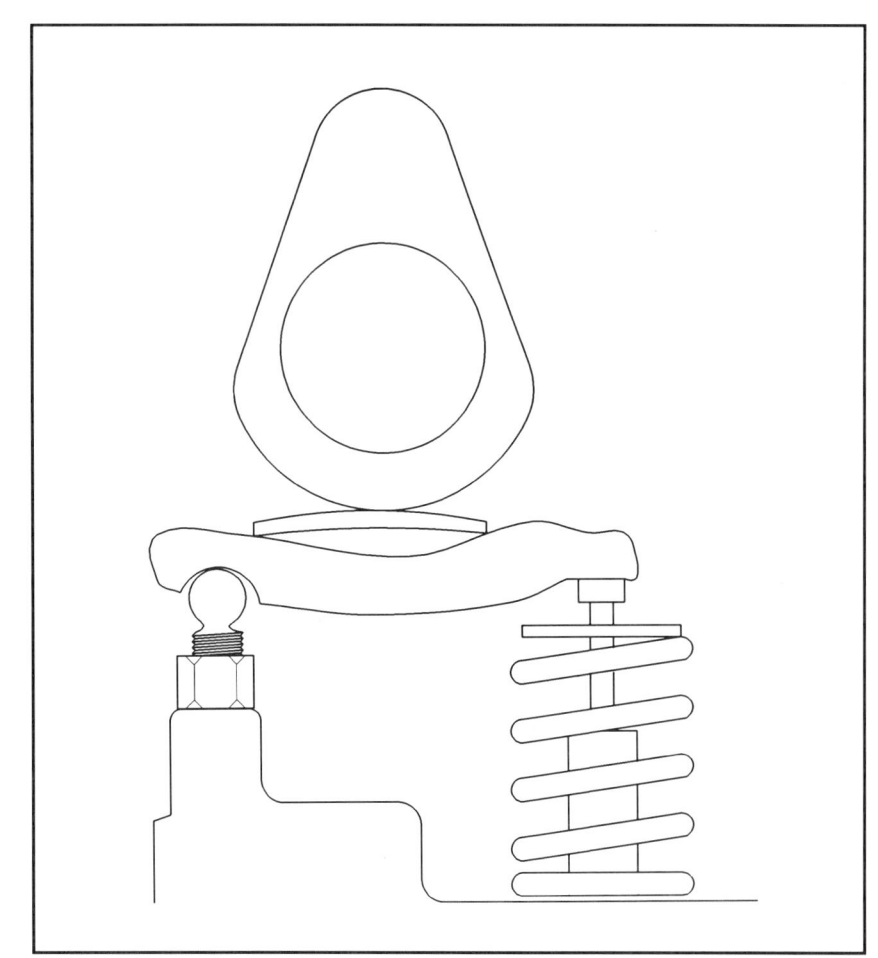

race from the planet Zaphod uses valve keepers as a nutritional source. Being a generous race, they take only one valve keeper from every head rebuilt in the galaxy. Protect your valve keepers.

Inspecting the Valve Springs

Look for broken or cracked springs. Set them on a flat surface, side-by-side. All of the intake springs and all of the exhaust springs should be the same height. If they are not the same height, replace them.

Professionally I have seldom replaced the valve springs during a head rebuild. However, several years ago the boss's father's boat broke down in British Columbia. One of the employees volunteered to drive the father's Cadillac to fetch parts and make the repairs. On the way, the Cadillac overheated severely. It was discovered the car had a severe misfire on two cylinders. Over a hundred dollars worth of ignition parts and a couple days of frustration later, we discovered that the overheating had weakened the valve springs. And as a trainer on the use of automotive diagnostic oscilloscopes, I have often seen scope patterns that indicate weak valve springs on heads with no perceivable driveability problems. Do not assume your valve springs are in good condition; have a machine shop test them or replace them.

Remove the Valves

Punch as many holes in a piece of cardboard as you have valves. Number the holes corresponding to the cylinders. Remove each valve, inspecting for wear on the valve stem and burning around the margin. Add to your parts list any that need to be replaced.

Although many technicians may argue the necessity of keeping the valves organized in the cardboard, at the very least it re-

The pivoting rocker overhead cam design has been used by manufacturers such as Mercedes-Benz. As the camshaft rotates the rocker is acted on directly by the camshaft. Adjustment of valve lash is typically accomplished in the design by altering the height of the pivot ball.

duces the possibility of them being damaged by being dropped.

Inspect and Measure the Valve Guides

Measuring the valve guides now will help you coordinate the work your machinist will be doing for you more effectively. Use a split ball gauge to measure the diameter at the top, just below the edge, at the bottom, and in the middle. If the valve guides are worn, the machinist will have several options for you.

First, if the guides are removeable, he can replace them. In fact, for those with a spirit more daring them the average mortal, you can procure a set of valve guide drifts, a press, and replace them yourself. If the guides are a cast part of the head they can be either knurled or sleeved.

Checking for Head Warpage

Place the cylinder heads on a bench. Check the head surface with a straightedge. As previously discussed, constantly changing temperatures make head warpage a common occurrence. Check for flatness along the longitudinal centerline of the block, as well as diagonally. Most heads, especially those made from the exotic alloys that import car manufacturers use, have characteristic warping patterns. These patterns often coincide with the places where head gaskets blow. Seek out the local expert on your particular engine. But beware: Many of the best sources for this information have a personality similar to a pit bull.

For short cylinder heads, such as V-8s and four-cylinders, the maximum allowable clearance between the straightedge and the head is 0.004in. For long cylinder heads on inline six-cylinder engines, the maximum allowable clearance is 0.006in. Carefully measure along the length of the cylinder head. There should be no more than 0.003in of warpage in any 6in length of head.

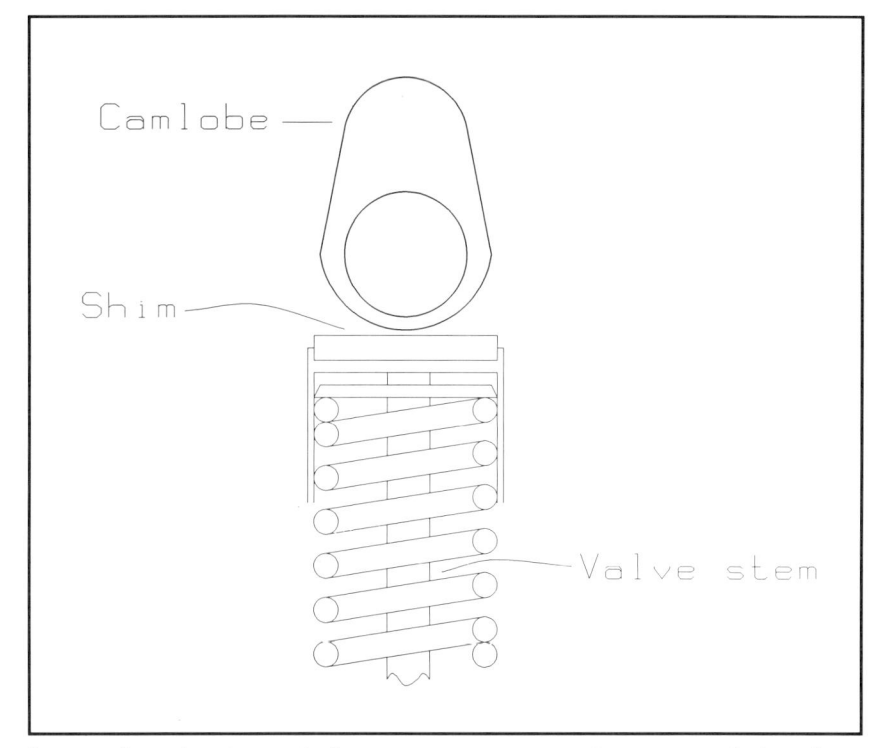

A second overhead cam design uses a cam follower. In effect the cam directly opens and closes the valve. This design is typical of that used by Volkswagen on its water-cooled engines. Adjustment is achieved by replacing shims on the top of the cam follower bucket.

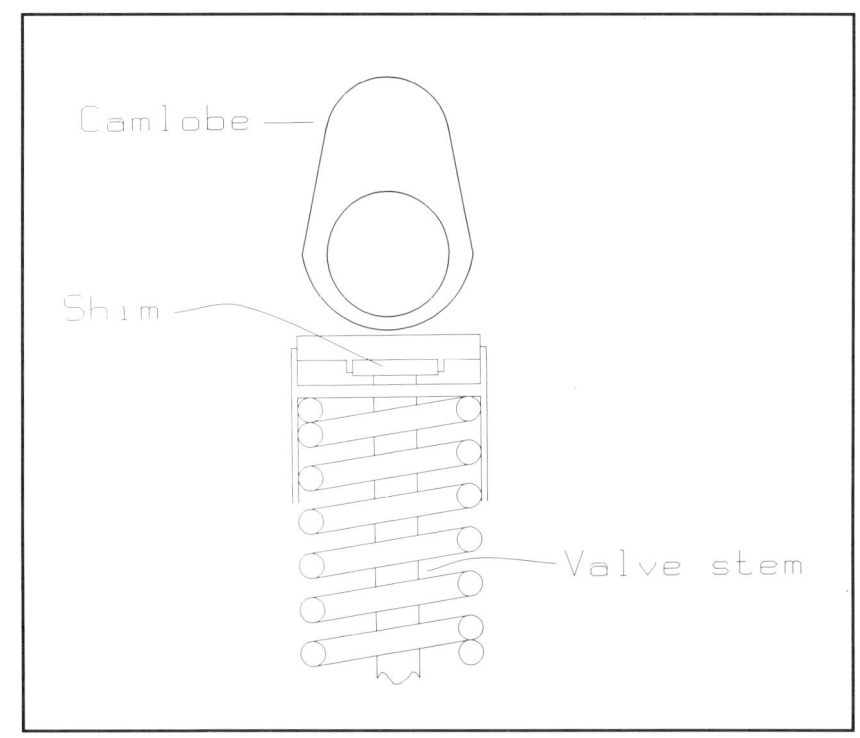

A third overhead cam design uses a cam follower also. In effect the cam directly opens and closes the valve. This design is typical of that used by SAAB. Adjustment is achieved by replacing shims under the cam follower bucket.

Obviously you will want to replace any broken valve springs. Also check the valve springs for cracking, height, and squareness. This valve spring is in fairly good condition but is not square. Lack of squareness in the spring may be evidence that the engine has overheated.

If the cylinder head in question is aluminum, inspect the area around the water ports of the head. If there is evidence of corrosion, the cylinder head must be resurfaced, even if it is flat.

Aligning Cam Journals

If the engine is overhead cam, the straightness of the top of the cylinder head is as important as the bottom. If the cylinder head has cam journals with little bearing caps resembling the main bearing caps of a block, remove the caps and check cam journal alignment with a straightedge. This technique is difficult, if not impossible to use on applications where the journals are one-piece in design. For these applications, slide the camshaft into place and rotate by hand. Since the most important consideration regarding the straightness of the top of the cylinder head is whether the cam will be put in a bind, this rather crude test is normally adequate. Again, as stated in the previous chapter, check the top of the block for flatness as well.

If the cam journals are not properly aligned, after the block mating surface of the cylinder head is machined flat, the cam journals will need to be align bored.

Measuring Valve Guides

The valve guides are measured using split-ball gauges. Slip the gauge into the guide. Since the guides generally are prone to both top-to-bottom wear and tapering, measure the guide at the top, in the middle, and at the bottom. The maximum allowable diameter is 0.003in greater than the diameter of the new valve. The maximum allowable taper is also 0.003in.

If the valve guides are worn or tapered, there are a couple of alternatives. First, the valve guide may be replaceable. Replacing the valve guides could amount to simply pressing the old guides out and the new ones in. For applications where the guides are part of the head, or nonreplaceable, the options are knurling or sleeving the guide. Sleeving the guide is an awkward, time-consuming, and expensive option. The procedure involves boring out the existing guide and pressing in a sleeve or insert. Knurling the guides is a much more efficient and more effective method.

Can you pick out the good valves and the bad valves in this picture? The bad valves have very thin margins and are clustered together just to the right of center in the picture.

Knurling uses a special tool to pull metal from the worn guide toward the center. This reduces the inside diameter of the valve guide to less than the diameter of the valve stem. A precision ream is then used to size the guide for the valve. While this may sound like a flaky repair, it is really an excellent way to repair the guides. Knurling leaves behind parallel grooves perpendicular to the centerline of the guide. These grooves help to control oil slippage down the guide, to be burned in the combustion chamber.

One symptom of bad valve guides is leaving behind a trail of smoke for your competitor to choke on as you decelerate into the next turn. While this may have some advantages in a racing venue, it is likely that your fellow commuters will mistake you for a city bus.

Checking Valve Stems

Since the valve stem is always made of a much harder metal or alloy than the guide, the stem will seldom wear significantly. To confirm that little wear has occurred, use a zero to 1in micrometer to measure each valve stem at three points.

Head Measurement Table

When you inspect and measure your heads, record the measurements so you can determine which components need to be machined or replaced. Make yourself a table with the following headings:
- Head Warpage
- Head Top Side Warpage
- Valve Guides Diameter
- Valve Guides Taper
- Valve Stem Taper

If there is too much wear or warpage on components, machine or replace those parts as necessary.

Checking Valve Springs

A complete check of the valve springs would include testing their tension. This requires special equipment and would need to be done by a machine shop. The tension of each spring should be within 5 to 10 percent of factory specification, and there should be no more than 10lb difference between the springs. If you do not have access to a spring tester and you are in doubt about the springs' condition, replace the valve springs. Assurance is always cheaper than doubt.

A quick and dirty test of the springs is a squareness and free-length test. Use a piece of glass and an accurate steel rule (a square would be best). Measure the springs one at a time. Rotate each spring as it sits on the glass. There should be less than 1/16in variation in the height of each spring as it is rotated, and less than 10 percent variation in free length between the springs. There should also be no more than 1/16in variation in the squareness of the spring.

The most common cause of variation in free length or squareness is overheating. All the springs were subjected to the same overheating condition. If

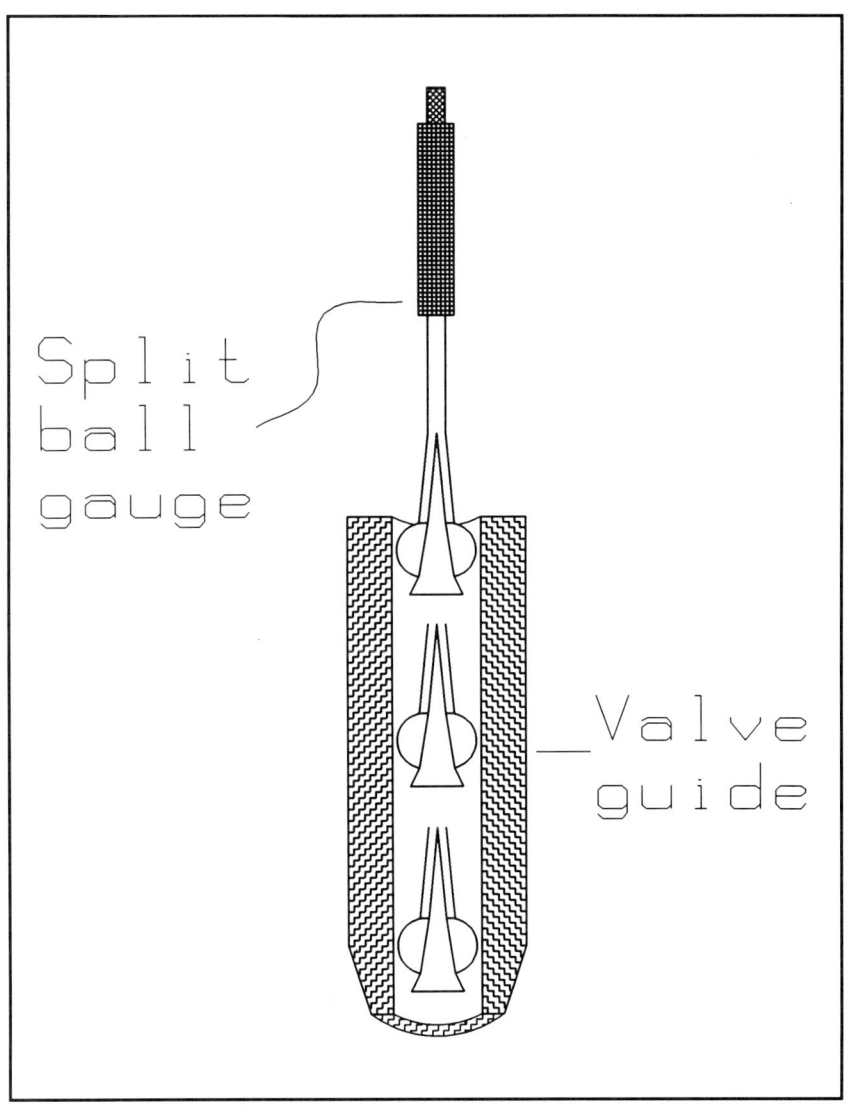

Small inside-diameter gauges, often referred to as split ball gauges, are used to measure the valve guides.

Measure the valve guide in three places.

75

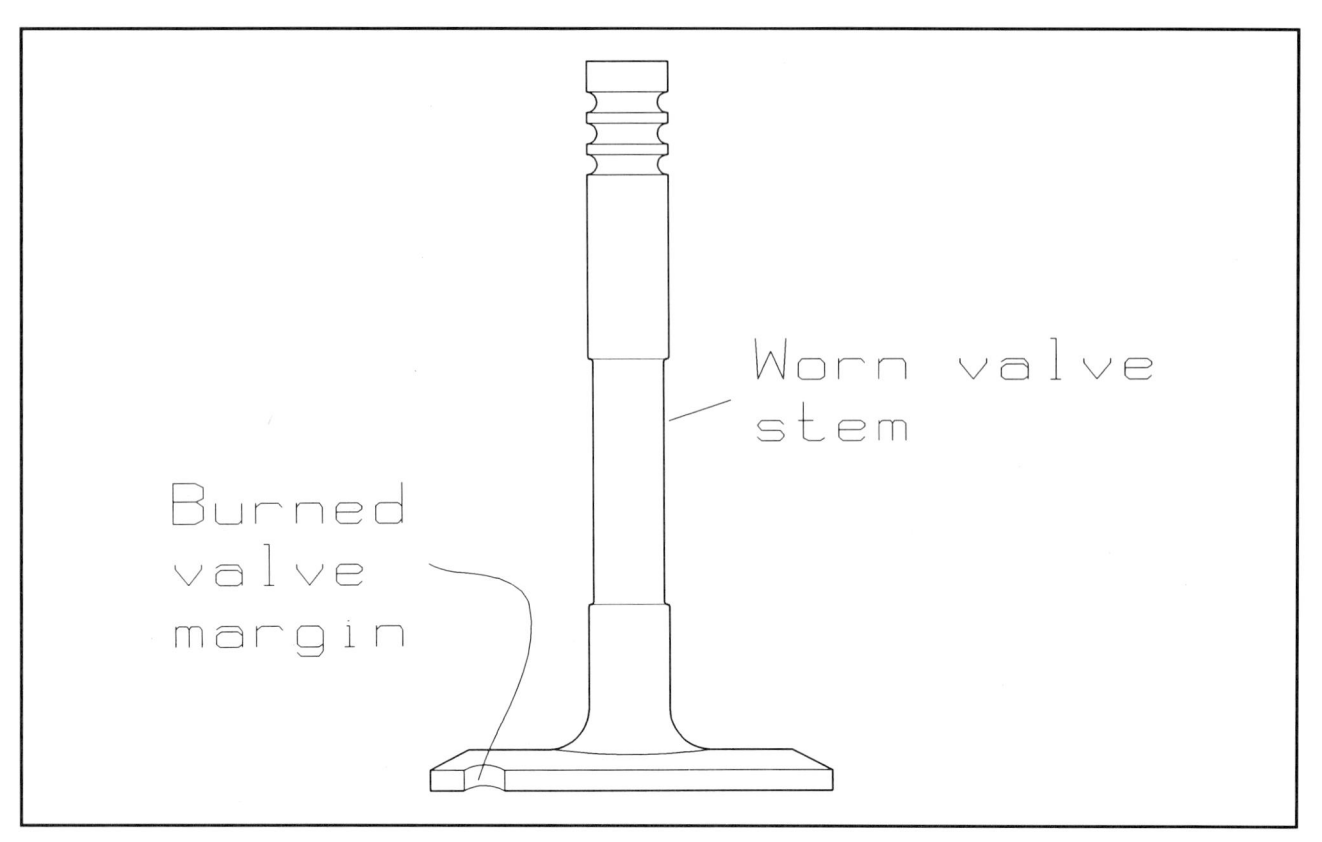

Measure the valve stem of each valve. Carefully compare the measurements at the top of the stem near the keepers and the middle of the stem where the valve rides in the guide.

Valve springs get very little attention among most "professional" technicians. I once wasted several days' labor, not to mention the great frustration, as a result of "tired" valve springs. If you are building up the heads for performance use or if the valve springs have been in the head for several years, replace them. High lift, long duration camshafts in particular require a valve spring that can close the valve quickly.

any of the springs show squareness of free-height problems, replace all the springs. Several years ago, the boss' father's boat broke down in British Columbia. One of the employees volunteered to drive the father's Cadillac to bring parts and make the repairs. On the way, the Cadillac overheated severely. Upon returning, the car had a severe misfire on two cylinders. Over a hundred dollars' worth of ignition parts and a couple of days of frustration later, we discovered that the overheating had weakened the valve springs.

Precision Measurements: Compression Ratio

In 1972, I formed a partnership with a college student to build a race car. As the "engine expert," I began to acquire Ford 302 parts. Having an incredibly limited budget, these parts were

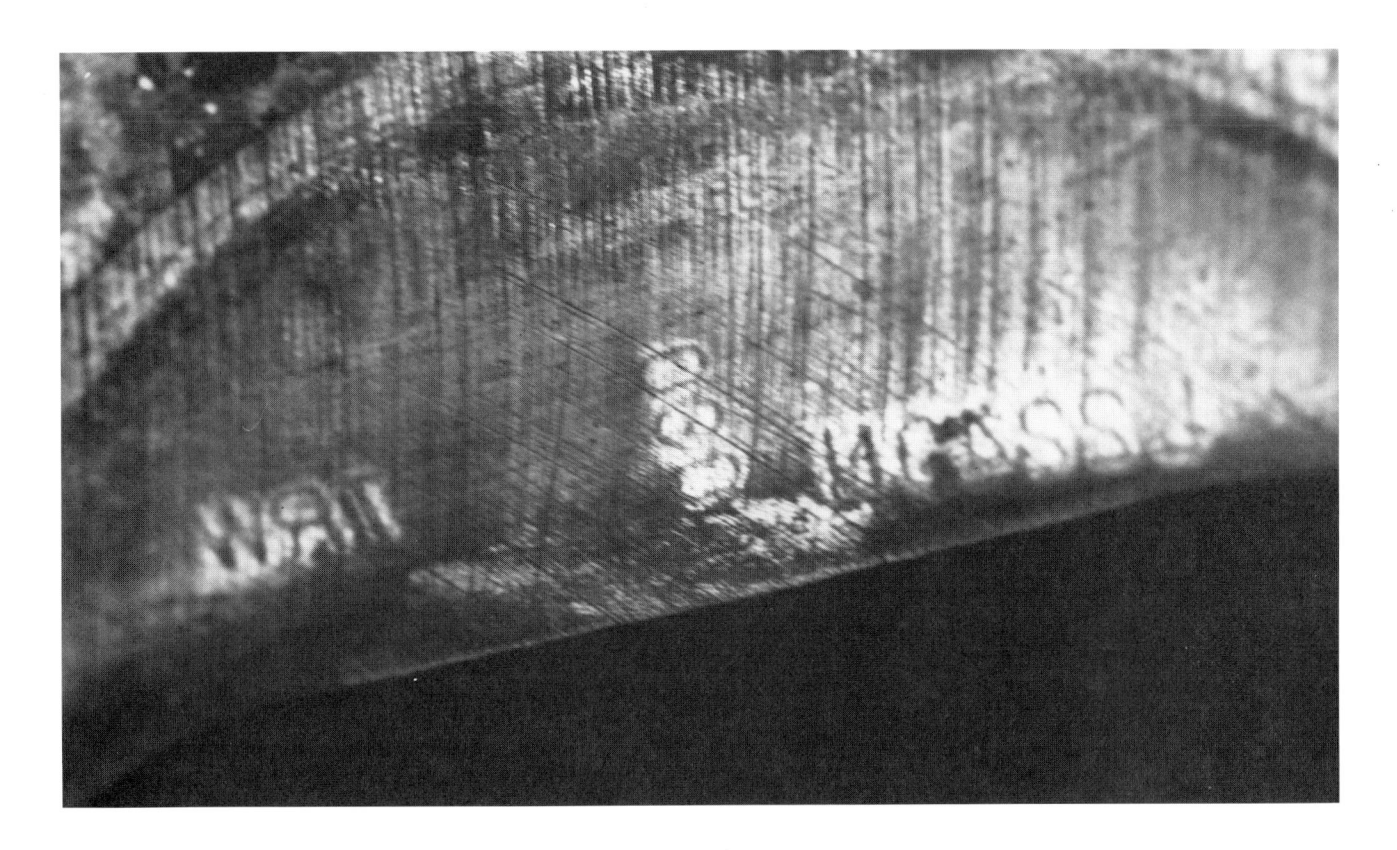

Failure to make accurate measurements can lead to catastrophic engine failure. Perhaps the most blatant example from my personal experience is illustrated by the bottom of this cylinder head (top photo). Notice that the markings are the same as the top of the piston (bottom). This is a result of trying to increase compression ratio without regard to consequences.

77

sourced through almost any legal means conceivable. The result was a very unusual collection of parts bolted together and dubbed "racing iron." After four seasons and two marriages (one each), the partnership dissolved and the engine sat on the floor of my partner's garage for more than fifteen years.

During the summer of 1992, I had the opportunity to tear down that engine and apply some of the knowledge gained during the intervening sixteen years. The first thing I noticed was that the heads were not from a 302, not even from a 289 or 260, but from a 221! The combustion chambers were extremely small, so immediately I wondered about the compression ratio. This is how I arrived at the correct figure:

Step 1: Measure cylinder head combustion chamber volume. With the valves in the head, set the head on a work bench at about a 45-degree angle. Place a bead of grease around the fringes of the combustion chamber. Since mineral oil will be used to measure the volume, be sure the grease being used will not immediately be dissolved by mineral oil. Place a piece of clear Plexiglass or Lexan over the combustion chamber, leaving a tiny crack at the top between the Plexiglass and the edge of the combustion chamber.

Using a graduated burette marked off in cubic centimeters (cc), slowly fill the combustion chamber with mineral oil. Subtract the level in the burette after the combustion chamber was filled from the level after the combustion chamber was filled. The difference is the cc volume of the combustion chamber. The size of the combustion chamber in the head is the total size of the combustion chamber—if the tops of the pistons are flat and they are flush with the block deck when they are at top dead center. If these two conditions are not true, proceed to step two. If they are true proceed to step three.

Step 2: Calculate piston volume of the combustion chamber. The piston can either increase or decrease the size of the combustion chamber.

In the mid-1970s, Environmental Protection Agency (EPA) concerns about emissions of oxides of nitrogen (NO_x) into the atmosphere caused car companies to take severe and even detrimental action. Since NO_x is formed in the engine when combustion temperatures exceed 2,500 degrees Fahrenheit, the manufacturers needed to decrease the combustion temperatures. The easiest way to do this was to lower the compression ratios. The easiest way to lower the compression ratios as to dish the tops of the pistons. This dishing increased the combustion chamber volume and thereby lowered the compression ratio.

Contrast this with my efforts on the 302 Mustang engine. I remembered that all the bench racers I had known in high school had emphasized the importance of high compression ratios, so I did as much as I could to up the compression. This included 12.5:1 compression pistons with rather impressive domes. These domes significantly filled the combustion chambers of the cylinder head.

Whether the shape of the top of the piston increases or decreases the size of the combus-

The first step to calculating compression ratio is to determine the size of the combustion chamber. This is done with a graduated burette.

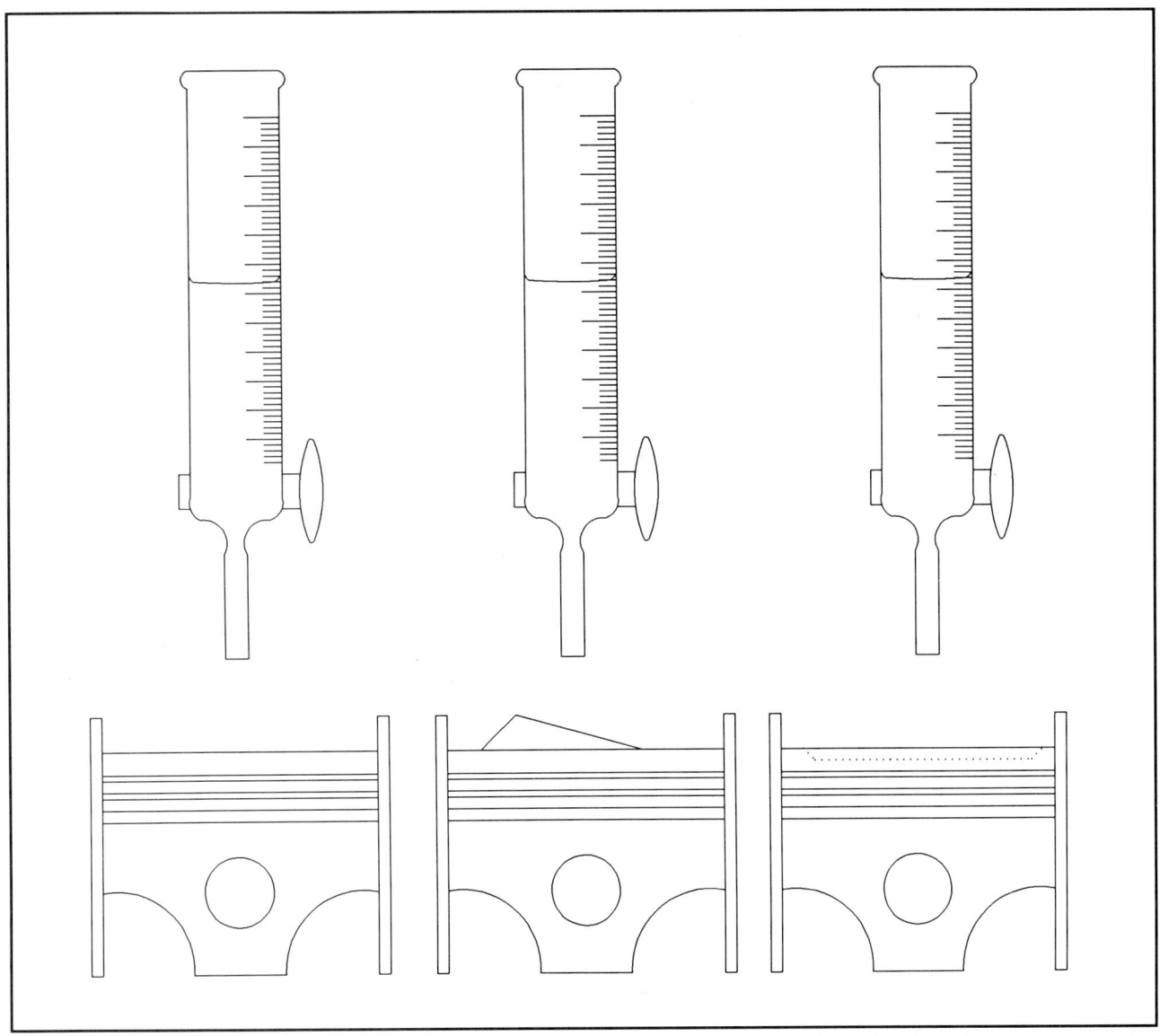

If the top of the piston is perfectly flat and level with the block deck, as illustrated in the example on the left, measuring the piston displacement of the combustion chamber is unnecessary. If the piston either increases of decreases the size of the combustion chamber, as in the two examples on the right, then the measurement is necessary.

tion chamber, the method of measuring its effect on the size of the combustion chamber is the same. Carefully position the piston 0.5in below the block deck. This can be accurately measured with a depth gauge or the end of a vernier caliper. With a straight edge ensure that the top of the piston does not protrude above the top of the block deck. Use the graduated burette to fill the area above the top of the piston. This measurement needs to be done accurately, but hastily; soon the mineral oil will begin to slip past the rings into the crankcase. Accuracy can be increased by carefully sealing the gap between the piston and the walls of the cylinder with grease before the procedure is begun. Once the measurement is made, the effect of the piston shape on the combustion chamber volume can be calculated.

Half-Inch Piston Displacement Measurement

If the piston top were flat, the volume of the cylinder above the piston less the amount of mineral oil poured out of the burette would equal zero. A domed piston would cause the amount of oil to be less than a flat-top piston. A dished-top piston would be greater than the flat-top. The difference between the flat-top ideal and the actual amount poured out

79

of the burette is the volume the piston adds or subtracts from the combustion chamber volume. The formula for this calculation is:

Cylinder Volume =
p / 4 x (bore x bore) x 0.5

This formula will give you the flat-top piston cylinder volume at 1/2in below the deck in cubic inches. Convert the results of this calculation to cubic centimeters using the following formula:

Cubic Centimeters =
Cubic Inches x 16.387064

The little Mustang engine of my youth sported a head chamber volume of 46cc. Since the cylinders were bored 0.30in over, the cylinder volume with a flat-top piston would have been:

Cylinder Volume =
(3.14/4) x (4.03 x 4.03) x 0.50
Cylinder Volume =
0.785 x 16.2409 x 0.50
Cylinder Volume = 6.37455325ci
Cylinder Volume =
6.37455325 x 16.387064cc
Cylinder Volume =
104.40602120792

When measured with a burette only 92cc of mineral oil filled the cylinder. This means that when the piston was at top dead center, it filled 12cc of the combustion chamber volume. The volume of the combustion chamber was therefore only 30cc.

Step 3: Calculate the cylinder volume with the piston at bottom dead center.

Cylinder Volume =
p/4 x (bore x bore) x stroke
Cylinder Volume =
p/4 x (4.03 x 4.03) x 3.0
Cylinder Volume =
0.785 x 16.2409 x 3.0
Cylinder Volume = 38.247ci
Cylinder Volume = 626.76cc

Since this is a domed piston application, the volume of the dome must be subtracted from the cylinder volume.

626.76 minus 12 = 614.76

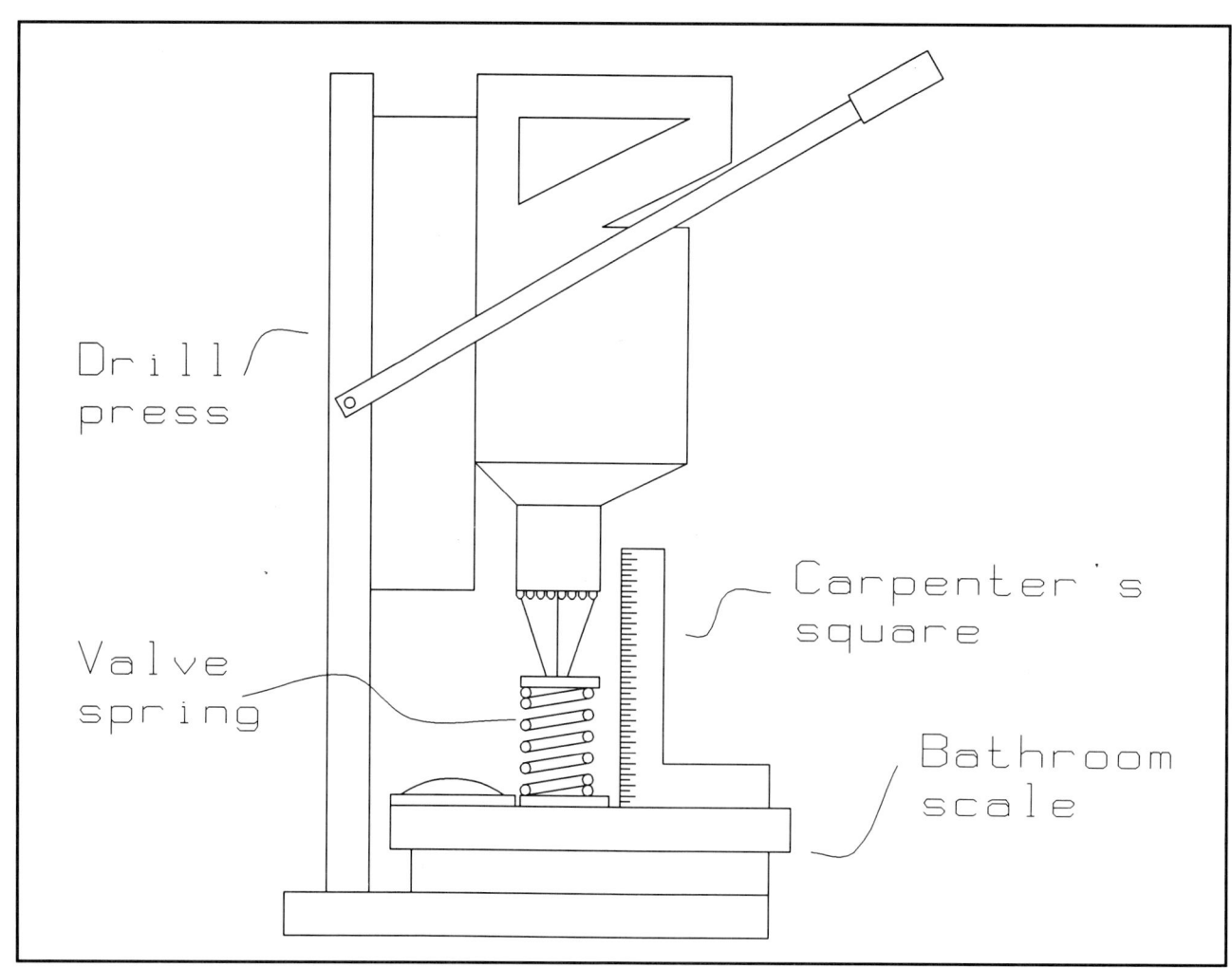

This little setup makes a dynamite way to check the valve spring tension. Watch the bathroom scale while you compress the spring to the rated height.

Step 4: Calculate the compression ratio.
Compression Ratio = (Chamber Volume + Cylinder Volume) / Chamber Volume
Compression Ratio = (30 + 614.76) / 30
Compression Ratio = 21.89:1

This is a bit more radical than I had intended. Even the fuels of the early 1970s were not acceptable for this kind of compression ratio; many diesels do not have compression ratios this high. The amazing part is that this engine actually started and ran. Of course, about 3500rpm is all I could depend on. Looking back, I am not surprised that I destroyed a starter at every race and had to replace the battery and the "00" gauge welding cable used as battery cables every season. Besides, the engine was only good for about ten 1-mile laps before it would blow a head gasket or two. Even with stock, flat-top pistons this engine-head combination would have netted a compression ratio of 14.6:1. This adventure is a perfect example of the traps that can be encountered when various engine parts are looked at individually rather than as a part of the entire package.

What are the problems associated with the compression ratio being this high? As the rising piston compresses the air to 1/16th its original volume, its temperature rises significantly to a point well above the ignition point of gasoline. This means that the fuel will pre-ignite during the compression stroke, limiting power and possibly damaging major engine components.

When a cylinder head is selected for an application, the size of its combustion chamber will have a major effect on the compression ratio. When fitted to the original 221 engine with flat-top pistons, the head just discussed gave the engine a compression ratio of 8.7:1 with the stock dished piston and 10:1 with a flat-top piston. The first thing you need to know when selecting the right head for the task intended is the piston displacement.

Let's go back to 1972: Gerry Beckley, Dewey Bunnell, and Dan Peak made it through the desert on some horse with no name. Richard Nixon was still a good guy (well, *pretty* good), and you could still get gas pumped for you. In the hot-car world, a young Smokey Yunick wanna-be was trying to select the proper head for his latest engine creation.

First, he measured the bore of the engine: 4.030in.

Okay, this 302 had been bored 0.030in over. Since there was no confirmed evidence that the crank was original, he measured the stroke with a depth gauge: 3.0in.

Although most of us in the real world are not blessed with a drill press of this grandeur, even a relative cheapy will work well for checking valve spring tension.

81

That is a stock stroke. Now, he had to decide on a desired compression ratio. In 1972, pump gas still had an octane rating higher than that of most city water systems, therefore he chose a compression ratio of 12:1. Today, he probably would settle for 10:1. The volume of the cylinder that will be displaced by the piston, regardless of dome, flat, or dish, will be 28.49ci. This converts to 467cc.

Chamber Volume =
Cylinder Volume /
(Compress Ratio - 1.0)
Chamber Volume = 626 / (12 - 1)
Chamber Volume = 57cc

Now the pistons and cylinder heads must be selected together to ensure that when the piston is at top dead center, there will be 57cc of volume in the cylinder head and any part of the piston that contributes to the combustion chamber.

Oddly enough, this is almost the exact size of a 351 Windsor head used with a flat-top piston. Even accounting for the few cc's of valve relief that would be machined in these pistons, this head would yield a compression ratio very close to the target. Once the combination is purchased and precise compression ratio measurements are made in accordance with the procedure previously described, fine-tuning the compression ratio can be accomplished by milling the head.

The amount to mill is based on the displacement ratio. The displacement ratio is the cylinder volume divided by the chamber volume.

Those who managed to complete their bachelor's degree in engineering without using a calculator will notice, when comparing the formulas below, that the displacement ratio is always the compression ratio less 1.0. Those who, along with me, barely conquered algebra in high school will have to take my word for it.

Compression ratio =
(Cylinder Volume + Chamber Volume) / Chamber Volume
Displacement ratio =
Cylinder Volume / Chamber Volume

Since the desired compression ratio is 12:1, the desired displacement ratio is 11:1. Let's say that precision measurements determine that the actual compression ratio with this piston-head combination is 11.5:1. The actual displacement ratio would be 10.5:1.

Amount to Mill =
[(New Displacement Ratio - Old Displacement Ratio) /
(New Displacement Ratio x Old Displacement Ratio)] x Stroke
Amount to Mill =
[(12 - 11) / (12 x 11)] x 3
Amount to Mill = (1/132) x 3
Amount to Mill = 0.0227in

By milling the head a little over 0.020in, our young engine builder would have had a track-burning engine instead of an overgrown doorstop.

Measuring Port Volume

Many readers probably see the internal-combustion gasoline engine as a piece of precision equipment. Other readers are probably laughing at that statement. The truth is that production engines are often highly imprecise chunks of iron. The manufacturers vary in their levels of precision. Even heads that are supposedly identical can have port volumes that vary tremendously.

In an ideal world, the performance engine builder would be able to go into a cylinder head warehouse and measure the port volumes until he found the heads with the largest and most closely matched volumes. Measuring the port is easily accomplished with Z-argon n-band scanning-tunneling sonar resonometer. Okay, the burette will work too.

Once the valves are installed and held in place with the valve springs, fill the port from the burette and record the number of cubic centimeters. Do this for each of the intake and exhaust ports. This procedure is particularly important if the head is being chosen for a racing engine in a class that does not allow head modifications. Several heads would be measured, and the ones with the largest ports would be chosen.

In an ideal world, there also would be reference specifications for port volume. In this world, however, the most important thing is that the head with the largest, most evenly sized ports be chosen. For most uses, the ports can modified later with a grinder.

Checking Port Shape

Actually, the shape of the port has more to do with its ability to flow air than just its volume. Visual inspection and estimation is impractical for even the most experienced eye. The solution is latex rubber. This stuff is hard to come by through traditional auto supply sources.

As I write this section, I am sitting in a hotel room in Anchorage, Alaska. In the phone book are four local companies that can probably supply me with the stuff, or at least lead me in the right direction. If all else fails, kits can be ordered from Chicago Latex Products.

Mix the latex according to the instructions. (Be sure to save a little to mold over a wig stand so you can play Martin Landau in "Mission Impossible.") Spray a penetrating oil into the ports to prevent the rubber from sticking. A spray lecithin mixture such as the commercial product Pam also works well. Pour the mixture into each of the ports. Then allow the mixture to set for several hours. Once it has set, remove the valves and pull the rubber out through the valve opening. To ensure even airflow, these molds should be as similar in shape as

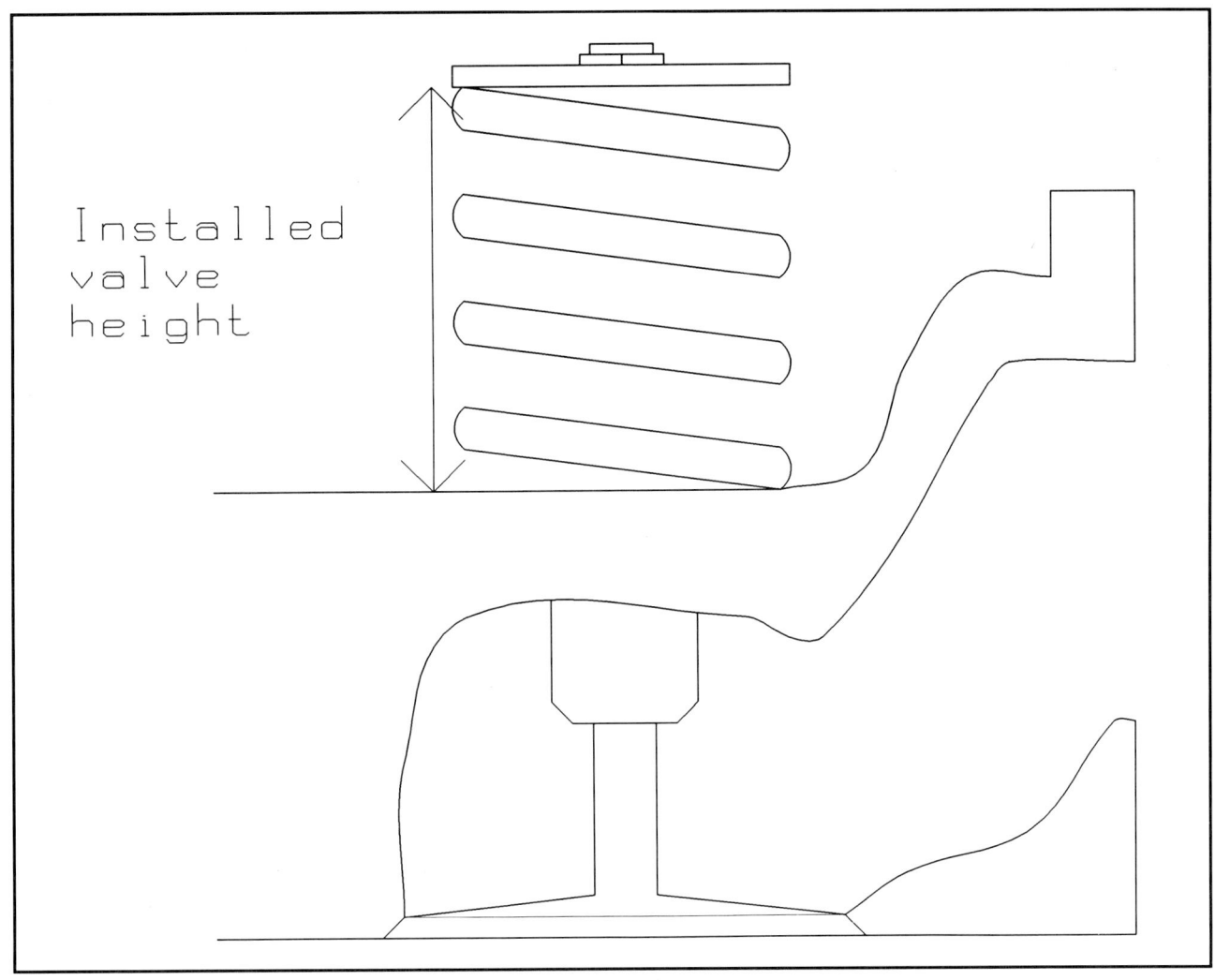

Installed valve spring height is the distance between the surface where the valve spring sits and the bottom of the valve spring retainer.

possible. Later in the book we will discuss how to use these initial mold and subsequent molds to port the head.

Measuring Valve Spring Tension

The first car I ever had was a Volkswagen Bug. With the engine in the rear, I quickly learned that the proper shift point was at the point of valve float. Fortunately, the engine had a lot of miles on it and the valve springs were weak, so valve float occurred well before maximum piston speed was attained.

Ignoring the tendency of pistons to disintegrate when their maximum rated velocity is exceeded, the single most limiting factor to engine rpm is valve float. This occurs when the cam's toe begins to lift the lifter or follower before the spring has had time to close the valve from the last opening. The greater the spring tension, the faster the valve will close. By the same token, the greater the spring tension, the faster the cam and lifters will self-destruct. The spring tension, therefore, is a compromise between the speed at which the valve will close, and the amount of punishment the rest of the valvetrain can take.

There is a tool specially designed to measure spring tension. One of the more well known manufacturers is Rimac. The valve spring is placed in the tool. A lever is moved to apply tension to the spring. The tension should be carefully measured at installed valve height and again at the spring height when the valve is fully open. The valve open height of the spring can be calculated by subtracting the net valve lift from the length of the valve, measuring from the seat to the bottom of the valve retainer. Compare these measurements to the specification provided by the engine or camshaft manufacturer. If either measurement is below spec, there are two alternatives.

The cheapest way to correct a below-spec spring is with hardened valve spring shims. Commonly available in thicknesses of 0.015, 0.030, and 0.060in, these shims can be added in combinations to get the tension specs up to where they belong. A better way to correct the problem is to replace the springs. After replacement, shims can be used to balance spring tension.

Note: At least one shim must be used under the spring of each valve on aluminum and alloy heads.

This is also a good time to check the valve springs for binding. Compress the valve spring to the valve's fully open height. Check this measurement carefully. Now compress the spring another 0.060in. If the spring is in a bind, either the spring must be replaced with one that has the correct tension and does not bind, or the net valve lift will have to be altered. For engines that are going to see extended service between overhauls, which really only excludes the most ardent and well-financed drag and sprint car racer, the figure of 0.060in should be adjusted up to 0.100in.

Not all of us are blessed with a Rimac or equivalent machine. A rather accurate and inexpensive valve spring tension tool can be constructed from a drill press and a set of bathroom scales. Place the scales on the platform of the drill press. Lay a piece of plywood over the scales, place the spring on the plywood, and use the drill press chuck to depress the spring. Control the pressure with the drill press lever. Keep in mind that if you are attempting to measure the valve's fully open tension, you will need the scales belonging to Sumo champion Akibono.

Measuring Installed Spring Height

A few paragraphs ago, I mentioned spring installed height and moved on as though I expected everyone to know how to measure it. Actually, it is quite simple to measure. Install the valve, the keeper, and the retainer in the head. It would be premature to make this measurement before the valve grind is done, as this will affect the measurement. Measure the distance between the bottom of the valve retainer and the surface where the valve spring sits. A telescoping gauge is perfect for this measurement; however, a dial indicator or even a steel rule will work well (with descending degrees of accuracy).

These are all the measurements that can be obtained before the head in installed. There will be more, once installation is complete. Unfortunately, though, reality can be cruel. All the work that has gone into precision measuring may be negated by one measurement at the time of reassembly.

Valves, Valve Guides, Seats, & Springs 10

The valves of a cylinder head typically are subjected to severe punishment. Visit a Volkswagen repair shop and you'll see why. Have one of the technicians put a used exhaust valve from an air-cooled engine in a vise. Put on a pair of goggles or safety glasses, and with a hammer strike the margin at right angles to the stem. It is likely that the valve will break. Then check the consistency of the metal on the inside of the valve. Often it will look more like concrete than metal. This change in consistency is the result of the extreme temperatures exerted on the exhaust valves of any engine. Air-cooled engines, especially, have to deal with high temperatures.

Today's valves are constructed in different ways and consist of varying materials in an effort to combat the effects of heat, and in an attempt to improve airflow and performance.

Types of Valves:
Steel Alloy

Steel alloy valves are the type used by most engine manufacturers. While they are wholly adequate for taking the family to the local grocery store, they leave a lot to be desired in performance applications.

The most common alloy used is a steel containing 21 percent chrome and 4 percent nickel. During rather tame engine operation, this alloy is both durable and inexpensive. Under the relatively light stress conditions put on the average commuter car, the heat generated in the combustion chamber is relatively low. As the piston travels down the combustion chamber prior to the opening of the intake valve, the hot post-combustion gases cool rapidly. Often by the time the exhaust valve opens, the temperature of these gases is below 1,000 degrees Fahrenheit. However, when the family teenager takes the car for a spin, these temperatures can climb rapidly.

Racing can create conditions where the temperature of the exhaust gases passing across the exhaust valve is well above 1,400 degrees F. At this temperature, 21C-4N steel alloy has a tensile strength one-sixth of what it is at room temperature. At 5000rpm, the valve springs slam the valve face against the valve seat 2,500 times per minute from the top end of the valve stem. The effect is a lot like walking a Great Dane that has a mind of its own. The dog pulls in one direction, while the dog-walker pulls in the other direction. If the leash (valve stem) is not strong enough, it will break. Add to this analogy the fact that the harder the valve spring pulls (as engine speed increases), the weaker the stem becomes due to the heat.

Sodium

Some valves contain sodium. This is a highly dangerous element as it will ignite upon contact with moisture. In the valve, however, the sodium fills a cavity. As the valve moves up and down in the guide, the inertial mass of the sodium attempts to keep it in one place. The sodium stirs within the valve, transferring the heat in the valve from the hot bottom to the cooler top. For many applications, these valves are standard equipment.

Stainless Steel

A relatively cheap and much stronger alternative to the chrome-nickel valve is the stain-

This is a typical steel alloy exhaust valve. The exhaust valve is always the smallest of the two valves in the head.

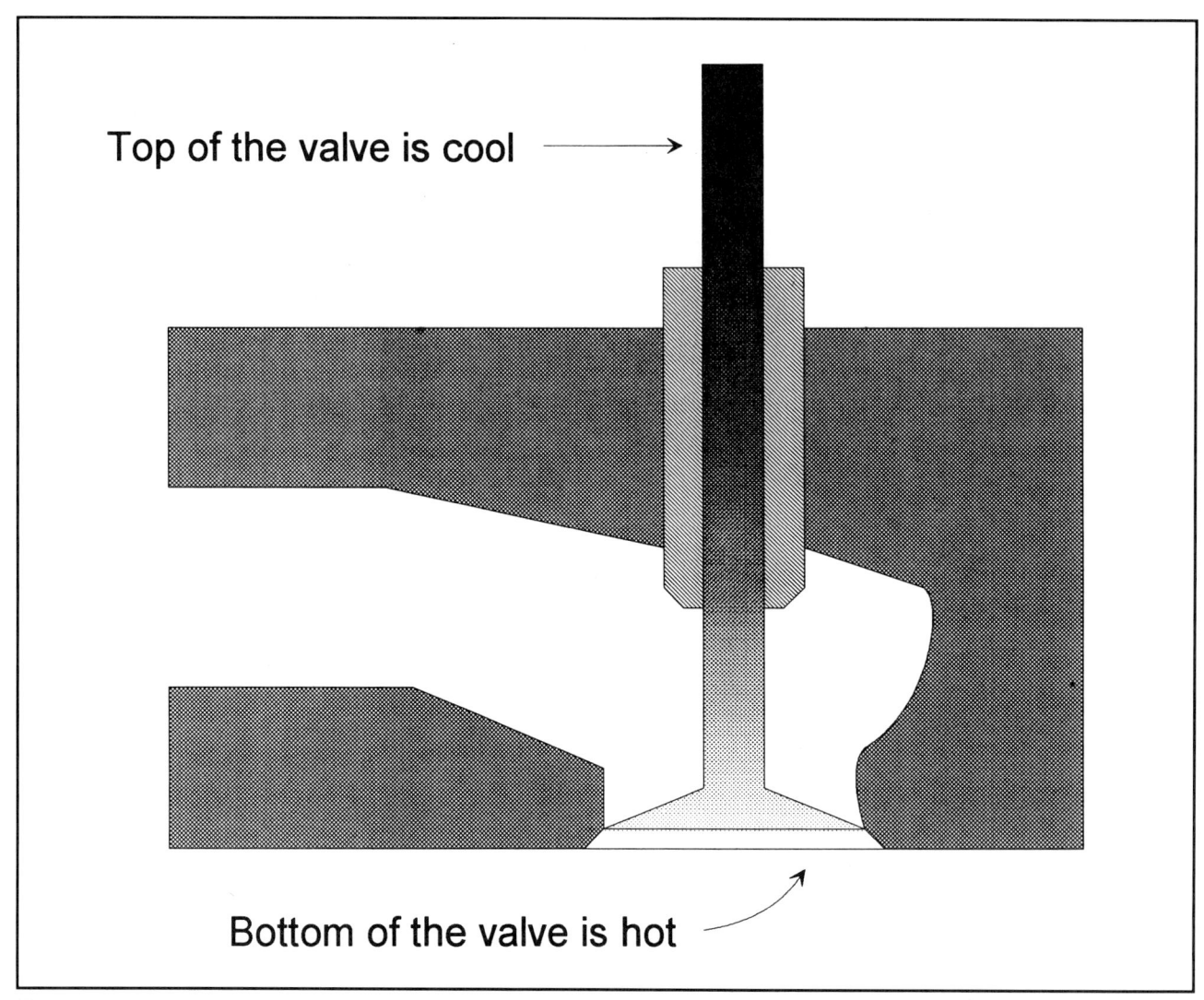

The temperature of the valve is vastly different from top to bottom. In an ideal valve the heat would be quickly and easily transferred from the bottom of the valve to the top.

less steel valve. For a typical domestic V-8 the valves will cost about $8-$20 each, compared to about $3-$5 each for stock valves. Now to my wife, $320 might be better spent on a pair of shoes for one of the kids. This attitude, of course, fails to recognize the *real* priorities in life. After all, I never had a pair of shoes until I was seven years old—and I had to wear *those* until I was eighteen.

Stainless steel valves have several advantages over stock valves. First, the cold tensile strength is greater than that of the stock valve. More important, the valve does not lose its tensile strength nearly as quickly when subjected to heat.

If high-horsepower high performance is your goal, paying extra for stainless will be worthwhile. Turbocharging or supercharging an engine, even one that will never see a track, can cause combustion temperatures to rise significantly. Strongly consider replacing the stock valves when moving away from normally aspirated engines.

For domestic V-8 engines, stainless steel valves are readily available at your local performance parts store. For those interested in more exotic wheels, say, an Alfa Romeo, these valves are going to be difficult to find. Stainless steel will be impossible to find for some applications, and therefore may have to be custom machined. In that case, be prepared to kiss Ben Franklin goodbye many times.

Titanium

When Lockheed was building the SR-71 spyplane, they chose a material that was virtually impervious to heat for the skin of the aircraft. Titanium alloys are extremely strong and do not lose their strength when heated. If the cylinder head is being built to push two tons of iron through 100 yards of extremely soft mud in less than five seconds, then titanium valves are ideal. Additionally, titanium weighs only about 70 percent as much as a stock valve. With less inertial mass, lighter valve springs can be used while still maximizing the rpm at which valve float occurs. The drawback? Incredible expense.

Ceramic

One of the latest technologies in engine components is ceramics. For most people, this term is wrought with memories of a cherished family heirloom that was destroyed during their inquisitive childhood. Light, strong, and virtually impervious to heat, these modern ceramics have been used in pistons, valves, connecting rods, and other engine components. While this use has been mostly experimental up to this point, the future looks bright for these materials. In fact, a Japanese company has used ceramic materials in an experimental engine that has no cooling system. The engine even has been designed to minimize the amount of heat loss to the air. With operating temperature over 500 degrees Fahrenheit, metal would distort. Ceramics undoubtedly will play a major role in the automobile engine of the future, but for today, cost and availability limit their use to high-budget custom applications.

Rotating the Existing Valves

Some applications attempt to decrease valve wear by rotating the valves. The standard valve spring retainer is replaced by a sprag-like device that turns the valve slightly as the valves are opened. While these are used occasionally on factory installations, they are not added to existing applications.

Valve Guides

A valve guide is more than just a hole in the cylinder head. The valve guide ensures proper alignment and lubrication of the valve stem. There are many types of valve guides available. On many cylinder heads, the valve guides are cast as part of the head while on others, the guides are pressed into the head.

Valve guides should be repaired or replaced anytime the cylinder head is worked. This is especially true if there was blue smoke coming from the tailpipe while the engine was decelerating. As mentioned earlier, the classic symptom of bad or worn valve guides is puffs of smoke during deceleration. The pressure in the intake manifold decreases dramatically (the vacuum increases). The higher vacuum causes the empty combustion chamber to seek air from anywhere. Excessive clearance between the valve stem and the guide allows the combustion chamber vacuum to draw air

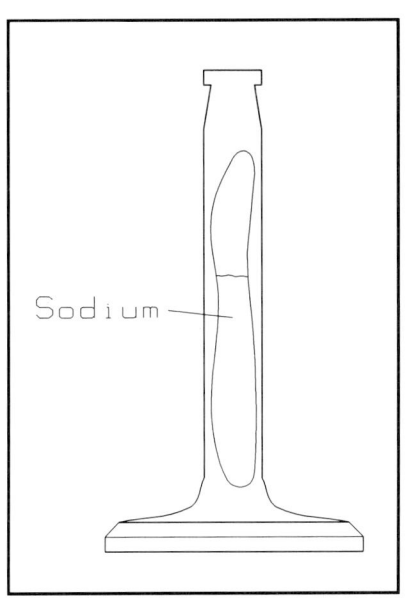

Many engines will use a sodium filled valve. Imagine a jar with a lid on it half filled with water. Shake the jar up and down like the valve moving in the guide. The sodium moves up and down in the valve and transfers the heat.

from the top of the cylinder head, beneath the valve cover. This air is saturated with oil, which is

Extremely popular today among racers and racer wannabes is the stainless steel valve. They are strong and permit smooth air flow across their surface.

drawn in with the air and burned in the combustion chamber.

Inspecting and Measuring Valve Guides

Measuring the valve guides now will help you coordinate more effectively the work your machinist will being doing. Use a split-ball gauge or a small-bore dial indicator to measure the diameter. As mentioned, since the guides are prone to both general top-to-bottom wear and tapering, measure the guide at the top, in the middle, and at the bottom. The maximum allowable diameter is 0.003in greater than the diameter of the new valve. The maximum allowable taper is 0.003in as well.

If the valve guides are worn or tapered, there are a couple of alternatives. First, the valve guide may be replaceable. Replacing the guides amounts to simply pressing the old guides out and the new ones in. Although it would be advisable to have these guides replaced by a qualified machinist during a valve grind, you can replace them yourself with a special set of drifts. These drifts not only have a surface for pressing the guide into place, but also have an undercut area that fits into the guide to prevent it from collapsing on itself.

Anytime the valve guides are replaced, a valve grind should be performed. For this reason, unless you plan on doing the valve grind yourself, just let the machinist replace the guides.

In spite of this, replacing the guides is a much easier operation than it might at first seem. The only special tools required are an appropriate valve guide drift, and a press or hammer. Slip the drift into the old guide and gently tap it out with a hammer, or better, use a press. Once the old guides are removed, install the new guides by tapping them into place. It is critical that the step of the valve guide drift be larger in diameter than the outside diameter of the guide. This is usually not a problem for drifts designed for this purpose. If the diameter of the drift is too small, it will damage the new guide.

For nonreplaceable applications where the guide is part of the head, the options are knurling or sleeving the guide. Sleeving the guide is an awkward, time-consuming, and expensive option. The procedure involves boring out the existing guide and pressing in a sleeve or insert. Knurling the guide is a much more efficient and more effective method.

Knurling involves using a special tool to pull metal from the worn guide toward the center. This reduces the inside diameter of the valve guide to less than that of the valve stem. A precision ream is then used to size the guide. While this may not sound like a high-quality repair, it is actually an excellent way to repair the guides. Knurl-

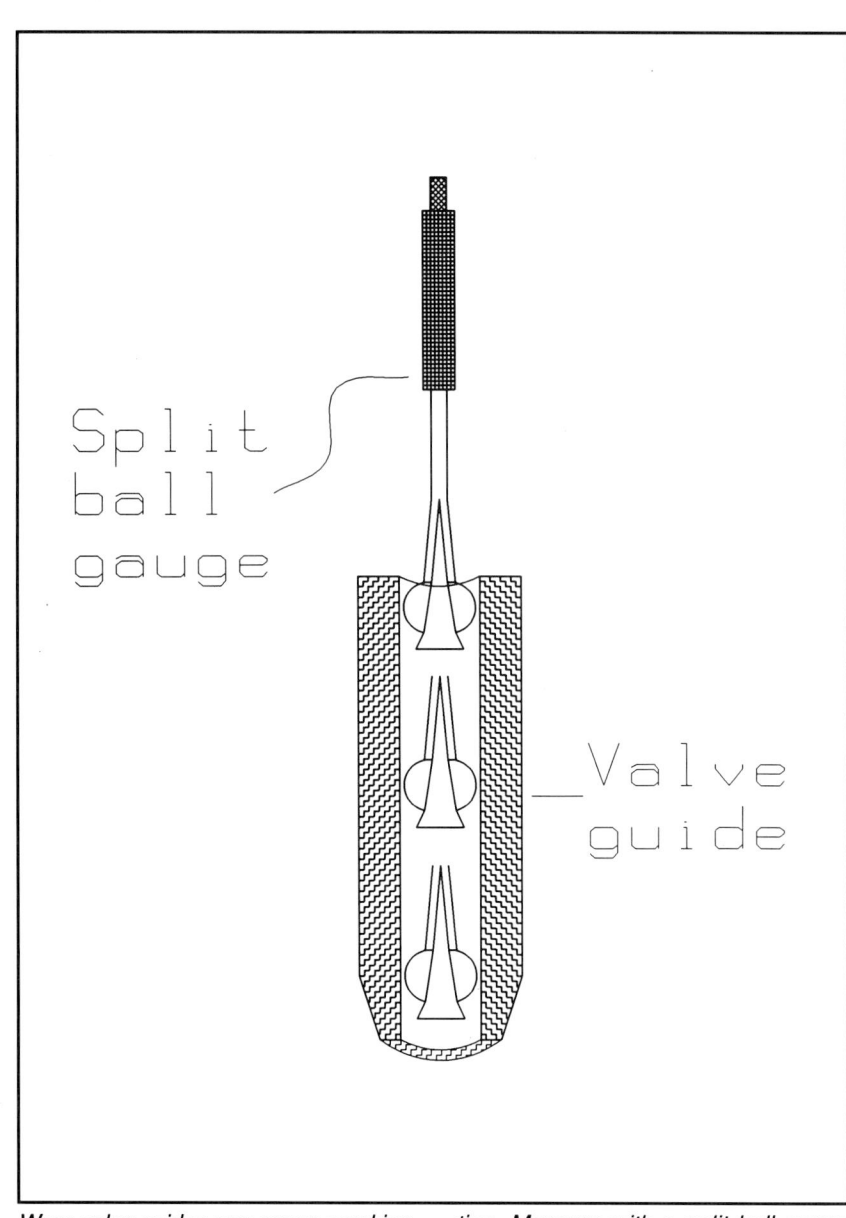

Worn valve guides can cause smoking on deceleration and erratic valve action. Measure with a split ball gauge and replace or machine as necessary.

ing leaves behind parallel grooves perpendicular to the centerline of the guide. These grooves help to control oil slippage down the guide, to be burned in the combustion chamber.

Although cast, the iron valve guides that are part of the cylinder head on most domestic engines work very well, and although knurling works effectively when guides are worn, old guides on performance engines should be repaired with bronze inserts. These inserts come in two forms, press-in and screw-in. Installation of both requires the skill of your friendly machinist.

Types of Replacement Guides

Most replaceable factory guides are made of cast iron. These guides are brittle and require a large clearance between the guide and the valve stem for proper lubrication. This clearance is 0.001 to 0.003in for the intake guides, and 0.002 to 0.004in for the exhaust. Excessive wear is considered anything greater than 0.005in.

The trick replacement for cast-iron guides is the phosphor bronze guide. These guides are porous in nature, and are self-lubricating. These qualities allow the guide-to-stem clearance to be very small. Although some machinists claim as little clearance as 0.0005in is acceptable with this type of guide, it is advisable to follow factory specs. In a performance application, the reduced clearance can provide better valve stability.

Once the guide is pressed into the head, it must be sized with a ream. The ream is designed with reverse flute, which will carry the bronze shavings up and out of the guide as it is reamed. A special lubricating oil should be used during this procedure, although many machinists advise that 90 weight oil works as well. Equally important as properly lubricating the ream, you must allow it to cool thoroughly between guides. In a professional machine shop the machinist will use two reams: one cools while the other is used.

Even bronze guides are limited by their unique metallurgical

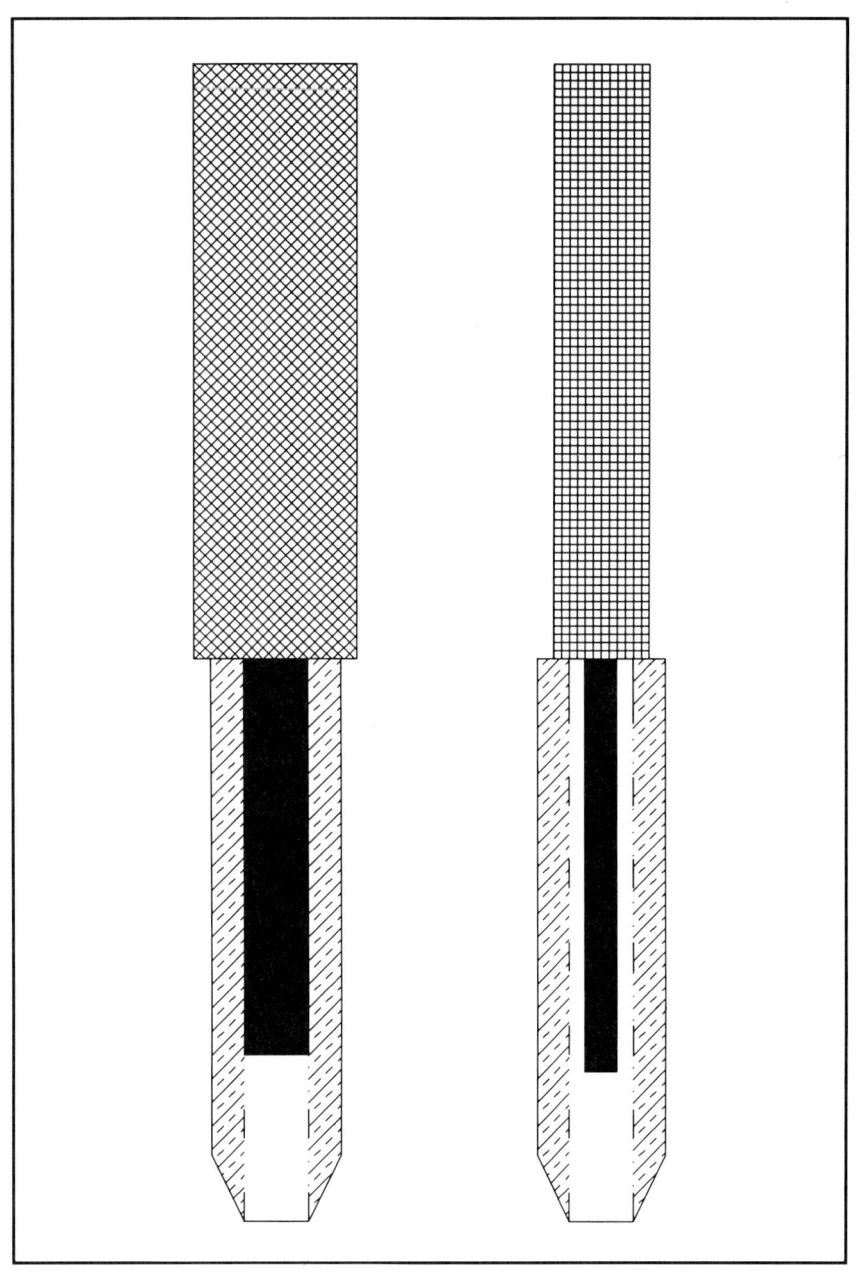

If the valve guides are removable, and many today are, they can be removed by pressing or driving them out of the head. In the tool chapter a set of these drifts was shown. Be sure to use a drift that is the appropriate size. If the wrong drift is used the old guide may flare and damage the head. If the wrong drift is used in installing the new guide, the new guide may be damaged.

makeup and design. Water-cooled Volkswagen products of the mid-1970s had a severe problem with worn guides. In one case, I personally experienced guides that were worn so badly that I could move the valve back and forth in the guide over an 1/8in with the

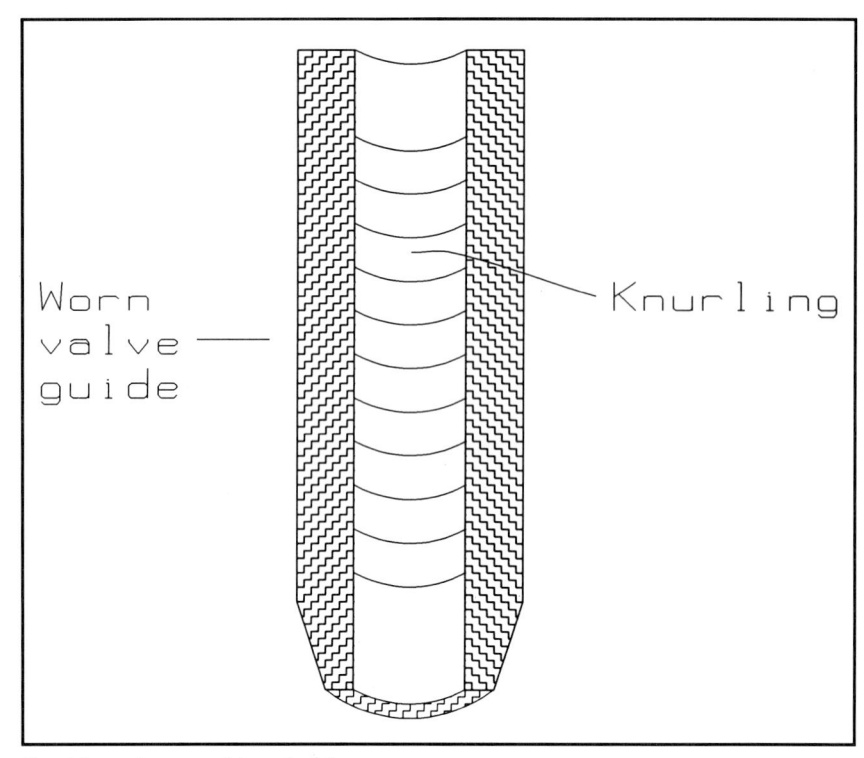

If guide replacement is out of the question, many machine shops can resize the guides by knurling them.

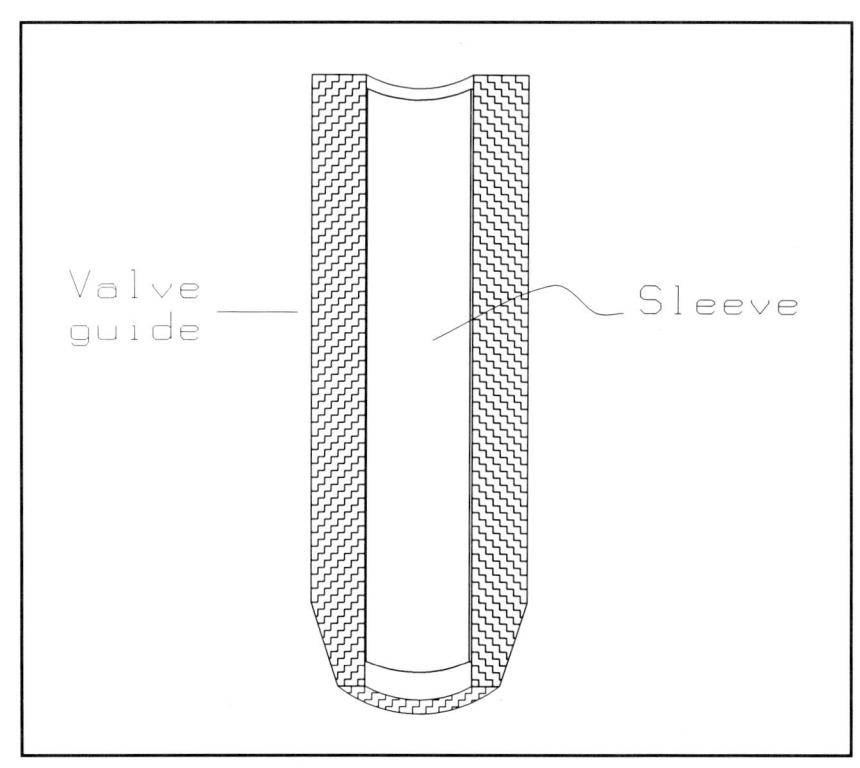

If the engine is destined for the race track, knurling is probably not advisable. A better alternative is a thinwall bronze insert. Almost any automotive machine shop can install these for you.

valve spring still installed. The problem was resolved by installing an updated guide.

Another unusual problem is when the guide comes loose in the head. Usually limited to more exotic engines such as Mercedes-Benz, this condition typically destroys the head. As the guide moves up and down in the head, it softens the metal around the guide. If the head is irreplaceable, a highly skilled machinist may be able to remove the damaged metal and either manufacture an oversize guide, or replacement weld the damaged area.

Valve Seals

There are as many kinds of valve seals on the market as there are poisonous insects in west Texas.

A popular valve seal on domestic engines is the umbrella seal. These just slip onto the valve stem, forming an umbrella over the top of the valve guide. The umbrella shape serves to shield the droplets of oil from the top of the valve guide, reducing the amount of oil that slips between the guide and the valve. Do not think, however, that new valve seals will cure oil burning resulting from worn valve guides. Valve guide repair is the only way to remedy that situation.

Some valve seals snap onto the guides and are held in place through friction. Many import car manufacturers prefer the snap-on type. Unlike the umbrella seal that tends to move up and down on the valve stem, friction holds the seal firmly on the guide. Being stationary, they have the additional benefit of wiping the valve clean of oil every time they travel down in the guide.

A word of caution: When selecting valve stems, make sure they are compatible with the valve springs you have selected. While this is rarely a problem with stock valve springs, many performance springs sport a larger diameter coil stock, or an inner

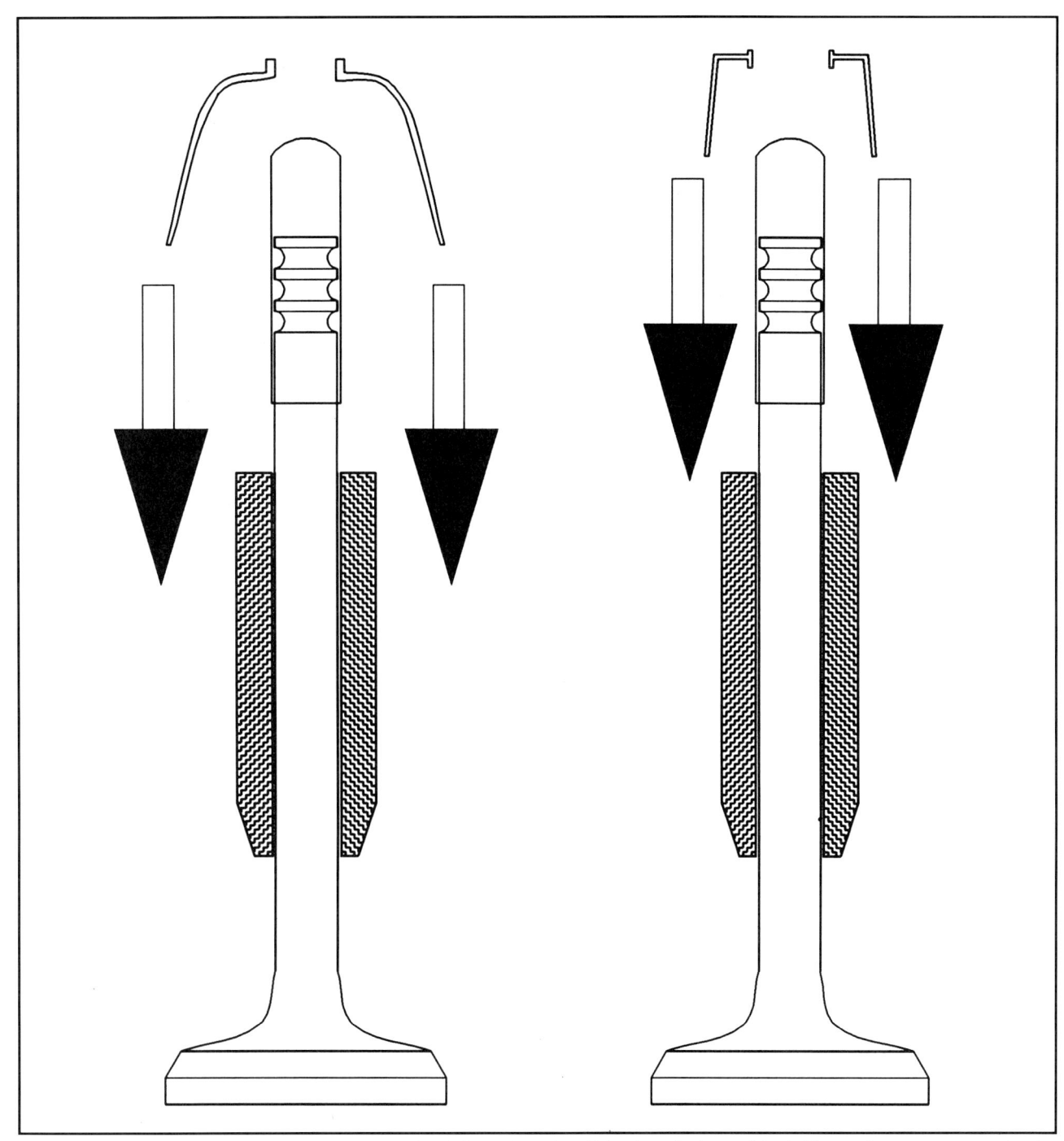

Valve seals are necessary to prevent oil from passing between the guide and the valve, especially during deceleration. The valve should be installed before the seal. With the valve seal kit a protective Mylar cover is usually provided. Place the cover over the installed valve, then install the seal.

spring. Often these springs can conflict with the valve stem seals. Select your valve springs first, then test the compatibility of the valve stem seal chosen. The clerk at your local performance parts house will be happy to cooperate; if not, his competitors will be.

Valve Springs

We have already discussed how to test valve springs. Now let us discuss how to select replacement springs. No parts work harder in the engine than the valve springs. To be honest, the only time I have ever tested valve spring tension was to ensure that

91

the tension of the *new* springs were balanced.

The basic concept here involves a race among engine components. The cam, lifters, and pushrods are in a race to open the valve before the spring can close it. The challenge for the engine builder is to balance the ability of the spring to rapidly close the valve without creating so much tension that it destroys the cam, lifters, and pushrods. Here are some things to keep in mind before purchasing new valve springs:
- Maximum rpm intended for the engine
- Lift of the camshaft
- Duration of the camshaft
- Net valve lift

As the speed of the engine increases, the cam, lifters, and pushrods have the advantage over the valve springs. Since science believes that it is impossible for two pieces of matter to occupy the same space at the same time, when the cam rotates and pushes the lifter or follower up, it *will* move up. Except for a little flex, when the lifter moves up, the pushrod *will* move up, the rocker *will* rock, and the valve *will* open. The valve spring has to deal with a far less definite set of conditions. Ideally, this spring would have no tension at all until just before the moment when the valve reaches maximum net valve lift. At that point, its tension would increase steadily to keep the slack out of the valvetrain. Since springs like this do not really exist, the engine builder must find a spring that will have just the right tension to keep the slack out of the valvetrain at engine speed a little above maximum designed rpm.

At the same time, any excess tension in the spring will contribute to wear in the valvetrain.

Valve Springs Types: Constant Tension

These springs feature evenly spaced coils. These springs are found primarily on low-tech engines. For these springs to be able to close the valve when a high-lift cam is used at high speed, the tension of the spring has to be quite high. Unfortunately, the tension at the beginning of the lift is about the same as the tension as the lifter or follower reaches the toe of the cam. This constant heavy pressure can cause rapid wear to the cam lobes.

Progressive

Most valve springs today fall into this category. The distance between the coils decreases from top to bottom. As the rocker arm or cam lifter begins to depress the valve, the tension of the valve spring is slight. As the spring continues to be compressed, the tension increases. The result is a valve spring that gives maximum tension when it is needed most, yet it provides minimum wear on the ramps of the cam lobe.

Dual and Triple

Double and triple springs use an outter spring that applies light tension to the valve when the valve is closed. The inner spring has a much greater tension. When the valve begins to open and the lifter has already begun to travel up the ramp of the cam lobe, the greater tension of the inner valve spring begins to come into play. This spring has a much greater tension than the outer. As the lifter or follower rides over the toe of the cam lobe, the high tension of the inner spring forces the closing of the valve to follow the lobe of the cam. As the lifter or follower nears the base circle of the cam lobe, the weaker outer spring finishes closing the valve.

The most common valve seals are the umbrella type and the snap-fit type. The umbrella type rides up and down over the guide with the valve stem, protecting the guide from oil rain. The snap fit type stays in place on the valve guide and seals out the oil.

Blueprinting

11

Back when I was in high school there was a fellow student whose father was the president of an oil company. The kid showed up the first day of our senior year with a Chevelle SS 396. Well we were all impressed (read that jealous). Rumors soon spread that this was not just any Chevelle SS 396, this engine had been balanced and blueprinted. When I heard that I nodded along with rest of the bench racing gang. We all nodded as though we knew what that meant. We were all impressed (read that jealous). As the years passed by the term fascinated me. Did it mean that someone had taken the trouble to make detailed drawings of every component in the engine? As more years passed by I found that defining the concept of blueprinting is a lot like defining what makes a good film, difficult to do precisely, easy to do generally. In a broad sense every engine built is blueprinted and balanced. When this term is used in the performance sense, the distinction is a matter of precision. Selecting a machinist to do your blueprinting is more difficult than selecting a good general automotive machinist to help you rebuild your engine. You want the kind of guy that measures from the top of the "i" to the dot when he signs a check. Precision is the order of business here.

Performance selections

Imagine the performance machine shop as a buffet restaurant. There are more performance modification options to a cylinder head than the typical engine rebuilder, or even performance modifier would want to pay for. Let's review the options.

Disassembly

Begin the disassembly by removing the valve cover. Inspect the valve cover for cracks, damage and warpage. If necessary, start a list of additional parts with a new valve cover. Remove the rocker arms or rocker arm assembly.

Inspecting Rocker Arms

On overhead valve engines there are three common rocker arm designs.

Rocker Design 1: The first design is a stamped steel rocker arm that mounts on studs protruding vertically from the cylinder head. The rocker arm is held in place by a nut. To facilitate rocking the rocker arm is held in place with a nut with a rounded bottom.

When removing these rocker arms check them for cracks in the area where the ball nut makes contact. In spite of their low-cost design this rocker arm design has few failures. The most common failure is when one of the nuts allows the tension created by the valve spring and the pushrod to force its way through the floor of the rocker arm. The second most common failure is when the pushrod is forced through the end of the rocker arm. Inspect this mating surface equally well. Wear can also occur on the machined surface where the rocker makes contact with the end of the valve stem.

Even a pro occasionally misses a deteriorated rocker arm. If there is any doubt in your mind about the condition of the rocker arms, add them to your list of things to buy.

Rocker Design 2: The second rocker arm has a shaft that runs from one end of the cylinder head to the other. This design features a cast rocker and an oiled bearing surface between the rocker arm and a shaft. The rocking surface of this type of rocker arm is pressure lubricated and therefore tends to be more dependable that the ball nut-mounted arm.

During disassembly inspect both the bearing surface of the rocker arm and the area of the shaft where that bearing surface mates. If either the shaft or any of the rocker bearing surfaces show any signs of wear replace the shaft and all the rockers. Also inspect the surface where the rocker contacts the top of the valve stem. If the shaft is in good condition, replace only the defective rocker. If more than one third of the rockers show significant wear on any of the wear surfaces, replace all rockers and the shaft.

Rocker Design 3: The third rocker arm design is much like the second. A single shaft supports and provides the pivot point for all of the rockers. Unlike the second design, however, this rocker has no actual bearing. This stamped steel rocker has a machined area which pivots on the shaft.

Check the surface which mates with the shaft for evidence of wear. Inspect the shaft. Make sure the push rod cup is in good condition as well as the rocker arm surface that comes in contact with the valve stem. If the shaft shows evidence of wear, replace the shaft and all the rockers. If the shaft is in good condition, replace only the worn rockers.

Pushrods

Find a piece of cardboard and punch two holes for each cylinder in it. Remove the pushrods from the engine and push them through the cardboard, noting which cylinder it is from and whether it is exhaust or intake. After all the pushrods are removed, lay a clean piece of glass on a flat surface. One at a time roll the pushrods across the glass. Any that do not roll easily are bent and should be replaced.

Lifters

If the engine is a V-6 or V-8, remove the intake manifold. Remember that if the cylinder heads on these engines are to be resurfaced it may be necessary to machine the intake manifold. Failure to machine the manifold mating surface to the cylinder head may result in vacuum leaks.

Remove the lifters from the block. If the lifters are to be reused they must be kept in order. A small box, just big enough for the lifters to fit in two rows for V-6 and V-8 applications or a single row for in-line applications would be a perfect solution. Label the box left/right, front/back. During an overhaul do not replace the camshaft without replacing the lifters and do not replace the lifters without replacing the camshaft.

Overhead cam

Overhead cam engines come in two common designs with respect to the actuating of the valves. In the first design the camshaft sits to the side and applies pressure to rocker arms which open and close the valves. In the second design the camshaft sits immediately above the valves and opens and closes them by means way of a cup and shim.

Rocker Arm Style Overhead Cam

On many overhead cam engines it will be necessary to remove the timing cover and remove the timing chain before removing the cam shaft. Be sure and draw a diagram of the alignment of the timing chain and gears before removing the chain. The rocker arm overhead cam design of overhead cam valve train place the camshaft to the side of the valves. A rocker arm is supported by a ball-tipped stud on one end and by the end of the valve stem on the other. The cam contacts a flat place on the rocker arm. The downward pressure of the cam against the rocker pivots it on the ball stud forcing the valve open.

When removing inspect the machined surface on which the cam rides for evidence of wear and pitting. In many engine designs these machined surfaces are lubricated very poorly. Also check the ball cup surface that mates to the ball stud. Wear on the machined surface that mates to the valve stem is also common. It is a fairly safe bet that you will find enough damage to warrant replacement of these rockers. Take this opportunity to carefully inspect the camshaft lobes for wear, rust, pitting, and other damage.

Shim and Cup Style Overhead Cam

Another overhead cam configuration is the shim and cup. A cup fits over the end of the valve which is recessed in the head. An adjustment shim fits in the top of the cup. As the cam rotates it presses on the adjustment shim which forces the cup downward, opening the valve. Remove the camshaft and inspect. Remove each of the valve shims. In most situations these shims will be reusable, inspect and set them aside to reuse during assembly. Please note that the chances of the same shim fitting in the same place after reassembly are pretty slim.

Inspect the sides of the cup for damage. Check for damage where the cup slides into the head. Make sure that cup moves smoothly in the head. If it does not, be sure to tell the machinist when you have the valve work done on the head.

Removing Valves

Although, as previously implied, there are ways of removing the valves that do not require a valve spring compressor, safety demands that I advise the use of one. Compress the valve spring with the compressor. Carefully remove the valve keepers and put them in a safe place. It is a little known fact of engine rebuilding that an extraterrestrial race from the planet Zaphod uses valve keepers as a nutritional source. Being a generous race they take only one valve keeper from every engine rebuilt in the galaxy. Protect your valve keepers.

Inspecting the Valve Springs

Look for broken or cracked springs. Set them on a flat surface side by side. All of the intake springs and all of the exhaust springs should be the same height. If they are not the same height, replace them.

Professionally I have seldom replaced the valve springs during an engine rebuild. However, several years ago the boss's father's boat broke down in British Columbia. One of the employees volunteered to drive the father's Cadillac to bring parts and make the repairs. On the way the Cadillac overheated severely. Upon returning the car had a severe misfire on two cylinders. Over a hundred dollars worth of ignition parts and a couple of days of frustration later we discovered that the overheating had weakened the valve springs. As a trainer on the use of automotive diagnostic oscilloscopes I have often seen scope patterns that indicate weak valve springs on engines with no perceivable driveability problem. Do not assume your valve springs are in good condition, have a machine shop test them or replace them.

Removing the Valves

Punch as many holes in a piece of cardboard as you have valves. Number the holes corresponding to the cylinders. Remove each valve, inspecting for wear on the valve stem and burning around the margin. Add to your parts list any that need to be replaced.

Although many technicians may argue the necessity of keeping the valves organized in the cardboard, at very least it reduces the possibility of them being damaged by being dropped.

Inspect and Measure the Valve Guides

Measuring the valve guides now will help you coordinate the work your machinist will being doing for you more effectively. Use a split ball gauge to measure the diameter at the top, just below the edge, at the bottom and in the middle. If the valve guides are worn the machinist will have several options for you.

First, if the guides are removable he can replace them. In fact, for those with a spirit more daring than the average mortal, you can procure a set of valve guide drifts, a press, and replace them yourself. If the guides are a cast part of the head they can be either knurled or sleeved.

Knurling the valve guides involves a tool which pulls metal toward the center of the guide. The next step is to precisely machine a new diameter in the guide.

Sleeving the guide involves drilling the guide larger and pressing in a metal sleeve. This technique is time consuming and has largely been replace by the methods mentioned above.

Checking the Head for Warpage

After removing the valves from the head, take a few minutes and clean the old gasket material from the surface that mates with the block deck. Now is the time to check for warpage. Place a machined straight-edge on the machined head surface. Using a feeler gauge check for clearance between the head and the straight-edge. Now lay the straight-edge diagonally across the head and check for twist.

Cracks in the Head

Carefully inspect the cylinder head for any evidence of cracking or damage. Begin with the assumption that the head is cracked. If you cannot see any cracks in the head then you can be sure the cracks are hidden. In order to prove the head is not cracked have it pressure tested.

Homer's law: You will never see the crack that causes the problem.

Cracked heads can sometimes be repaired, ask your machinist.

Measuring the Cylinder Heads
Checking for Head Warpage

Pick up the cylinder heads and place them on the bench. Check the head surface with the straight-edge. Constantly changing temperatures makes head warpage a common occurrence. Check for flatness along the longitudinal centerline of the block as well as diagonally. For short cylinder heads, such as those of V-8s and four-cylinders, the maximum allowable clearance between the straight-edge and the head is 0.004in. For long cylinder heads on in-line six-cylinder engines the maximum allowable clearance is 0.006in. Carefully measure along the length of the cylinder head. There should be no more than 0.003in of warpage in any 6in length of the head.

If the cylinder head in question is made out of aluminum, inspect the area around the water ports of the head. If there is evidence of corrosion the cylinder head must be resurfaced even if it is flat.

Cam Journal Alignment

If the engine has an overhead cam the straightness of the top of the cylinder head is as important as the bottom. If the cylinder head has cam journals with little bearing caps resembling the main bearing caps of a block, remove the caps and check cam journal alignment the same way you tested main bearing bore alignment. The technique is difficult or impossible to use on applications when the overhead cam journals are one-piece in design. For these applications, slide the camshaft into place and rotate by hand. Since the most important consideration regarding the straightness of the top of the cylinder head is whether the cam will be put in a bind, this rather crude test is normally adequate.

If the cam journals are not true, after the block mating surface of the cylinder head is machined flat the cam journals will need to be align-bored.

Measuring the Valve Guides

The valve guides are measured using split ball gauges. Slip the gauge into the guide. Since the guides are prone to both general top to bottom wear and tapering, measure the guide at the top, in the middle and at the bottom. The maximum allowable diameter is 0.003 greater than the diameter of the new valve. The maximum allowable taper is 0.003.

If the valve guides are worn or tapered, there are a couple of alternatives. First, if the valve guide may be replaceable. Replacing the valve guides would amount to simply pressing the old guides out and the new one in. For applications where the guide is part of the head, non-replaceable, the options are knurling or sleeving the guide. Sleeving the guide is an awkward, time consuming and expensive option. The procedure involves boring out the existing guide and pressing in a sleeve or insert.

Much more efficient and even more effective is to knurl the guide.

Knurling involves using a special tool to pull metal from the worn guide toward the center. This reduces the inside diameter of the valve guide to less than the diameter of the valve stem. A precision ream is then used to size the guide for the valve. While this may sound like a flaky repair, in reality this is an excellent way repair the guides. Knurling leaves behind parallel grooves perpendicular to the centerline of the guide. These grooves help to control oil slipping down the guide to be burned in the combustion chamber.

A symptom of bad valve guides would be puffs of smoke in the face of the competition you just passed as you decelerate into the next turn.

The Valve Stems

Since the valve stem is always made of a much harder metal or alloy than the guide, the stem will seldom wear significantly. To confirm that little wear has occurred use a zero to 1in micrometer to measure each valve stem at three points.

Valve Springs

A complete check of the valve springs would include testing their tension. This requires special equipment and would need to be done by a machine shop. The tension of each spring should within 5 percent to 10 percent of factory spec and there should be no more than 10lb difference between them. If you do not have access to a spring tester and you are in doubt about their condition, replace the valve springs. Assurance is always cheaper than doubt.

A quick and dirty test of the valve springs is a squareness and free length test. Use the piece of glass you used earlier to check the pushrods and an accurate steel rule (a square would be best). Measure the springs one at a time. Rotate each spring as it sits on the glass. There should be less than 1/16in variation in the height of each spring as it is rotated and there should be less than 10 percent variation in free length between the springs.

The most common cause of variation in free length or squareness is overheating. All the

This valve grinding machine is the type that has been the standard of the industry for decades.

springs were subjected to the same overheating condition. If any of the springs show squareness or free height problems, replace all the springs.

The 'Performance' Valve Grind

My opinion of what is generally marketed as a performance valve grind was stated in an earlier chapter. This statement did not mean to imply that there is no such thing as a performance valve grind only that most performance valve grinds are not what they profess to be. When a typical "performance" valve grind is done the seats are cut at three angles. A 45 degree cut is used to serve as the mating surface with the valve.

If the machineist feels that the valve seat is too wide or mates with the valve in the wrong place, he will narrow or raise or lower the seat. To accomplish these things he will narrow the seat by grinding it at either a 30 degree or 60 degree angle. Use of a 30 degree grinding stone or cutter will also lower the seat. Use of the 60 degree stone will lower the seat.

This is where the typical "performance valve grind" stops. A far cry from Billy Ray grinding the valve on his 409 with a lapping stick a Elrod's Drive-in, but hardly a performance grind.

Grinding the Valves

"New valves, fresh out of the box do not need to be ground." "Government always has the best interests of the people at heart." "Children are never selfish." I suppose you get the point. Theoretically the valve should not need to be ground, however, the minor damage that occurs during shipping could limit the sealing ability of the new valve. Minor warps, face and margin damage can all occur during transportation. This damage is almost always minimal and usually would have no effect on the quality of a regular valve grind. Remember,

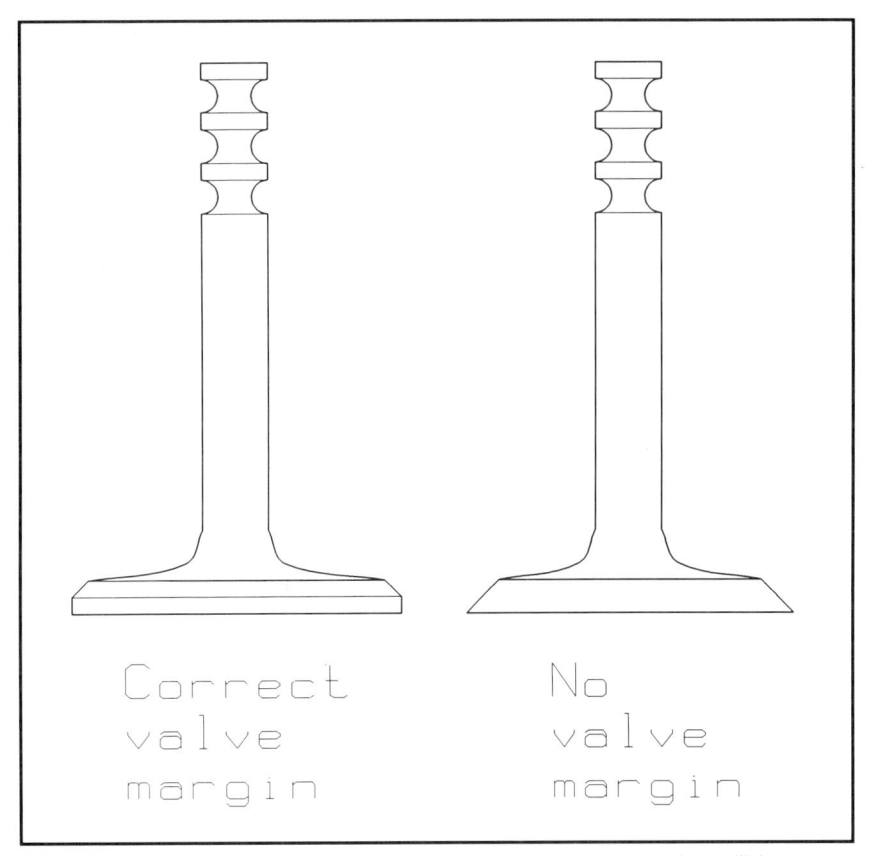

After the valve is resurfaced there should be a margin. If the valve is ground to a sharp edge the edge will overheat and the valve will burn very quickly.

however, that we are talking about a high performance grind, anything less than perfection is undesirable. Grind the valve to a 45 degree angle.

Note: Some valves have an aluminum coating to protect them from corrosion. Do not grind these valves before installation.

Valve Seat

Several years ago I made the fatal mistake of an independent shop owner, I worked on a friend's Volkswagen van. Actually this was two fatal mistakes, working on a friend's car and working on a Volkswagen van. The friend convinced me to do something I would never have done on a regular customer's car, I used remanufactured heads. Shortly after overhauling the engine I found myself taking a 500 mile round trip to rescue him and his marooned family. In less than 350 miles two valve seats dislodged themselves from the cylinder head. A valve seat hanging around the valve stem can make a very unusual set of noises. Most people who have never worked very much with air-cooled Volkswagens and remanufactured heads do not have a true appreciation of valve seats.

The valve seat has several jobs. The seat must provide a good seal with the valve, holding the combustion gasses in the cylinder. The valve seat is also responsible for carrying heat away from the valve. Additionally the valve seat must be durable enough to withstand the constant pounding which results from the closing of the valve. Each valve will be slammed closed on a V-8 between 250 and 2500 times per minute depending on engine

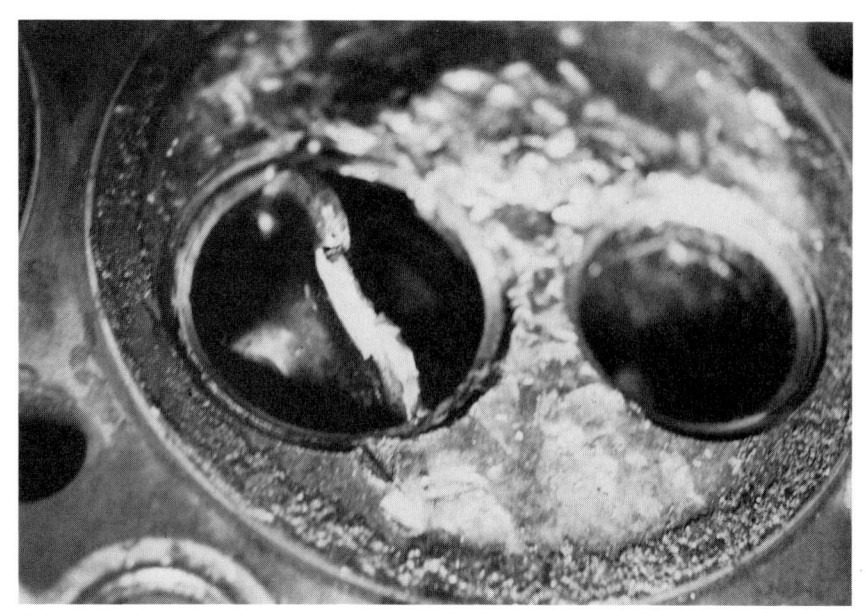

This is the end result of a catastrophic valve failure.

speed. During an eight hour vacation drive in the family station wagon each valve will close about 600,000 times. Each time the valve closes the valve spring may supply several hundred PSI of force to the mating of the valve and the valve seat. There are several types of valve seats.

Integral

For decades it was common practice for the valve seat to be part of the cylinder head. To this day they have the advantage of running 100 to 200 degrees cooler than replaceable seats. This meant that after repeated valve grinds or catastrophic damage to the seat it would be necessary to replace the head. Integral valve seats remain common practice on cast iron heads. However, cast iron heads do not remain common practice. In the eighties most new engine designs opted for alloy heads.

The advantage of integral valve seats is simplicity. The problem described above concerning the Volkswagen van could not have occurred in a cylinder head using integral seats. For applications where extremes of temperature or extremes of use are found the integral cast iron seat may not be adequate. Adding a turbocharger can increase intake manifold temperatures by dozens or even hundreds of degrees. This can increase the temperature of combustion and of the exhaust gasses by several hundred degrees. The cast iron seat may not be able to endure these temperatures. Another way of stressing the valve seats to their limit is by constant operation of the engine. The typical commuter car gets about two hours of operation per day or less. (Of course if the reader is currently sitting on I-405 between Long Beach and LAX he might disagree with that rather conservative estimate.) In a year this "typical" car may get 1,000 hours of operation. This is a very generous estimate of 35,000 miles. Most drivers will only incur 15,000 to 25,000 miles. That is an estimated 430 to 720 hours of operation. Now imagine an engine being used in a working boat (as opposed to pleasure boat) or in a primary generator. Conceivably these applications could get over 8000 hours of operation in a year, the equivalant to a quarter million miles a year. I consult with many state departments of transportation that have vehicles the operators will allow to idle eight hours a day. They idle to power air conditioners, arrow boards or miscellaneous equipment. In one month these vehicles experience a run time

For decades most domestic cylinder heads have incorporated integral valve seats. The valve seat is part of the casting of the cylinder head and cannot be removed without machining it out.

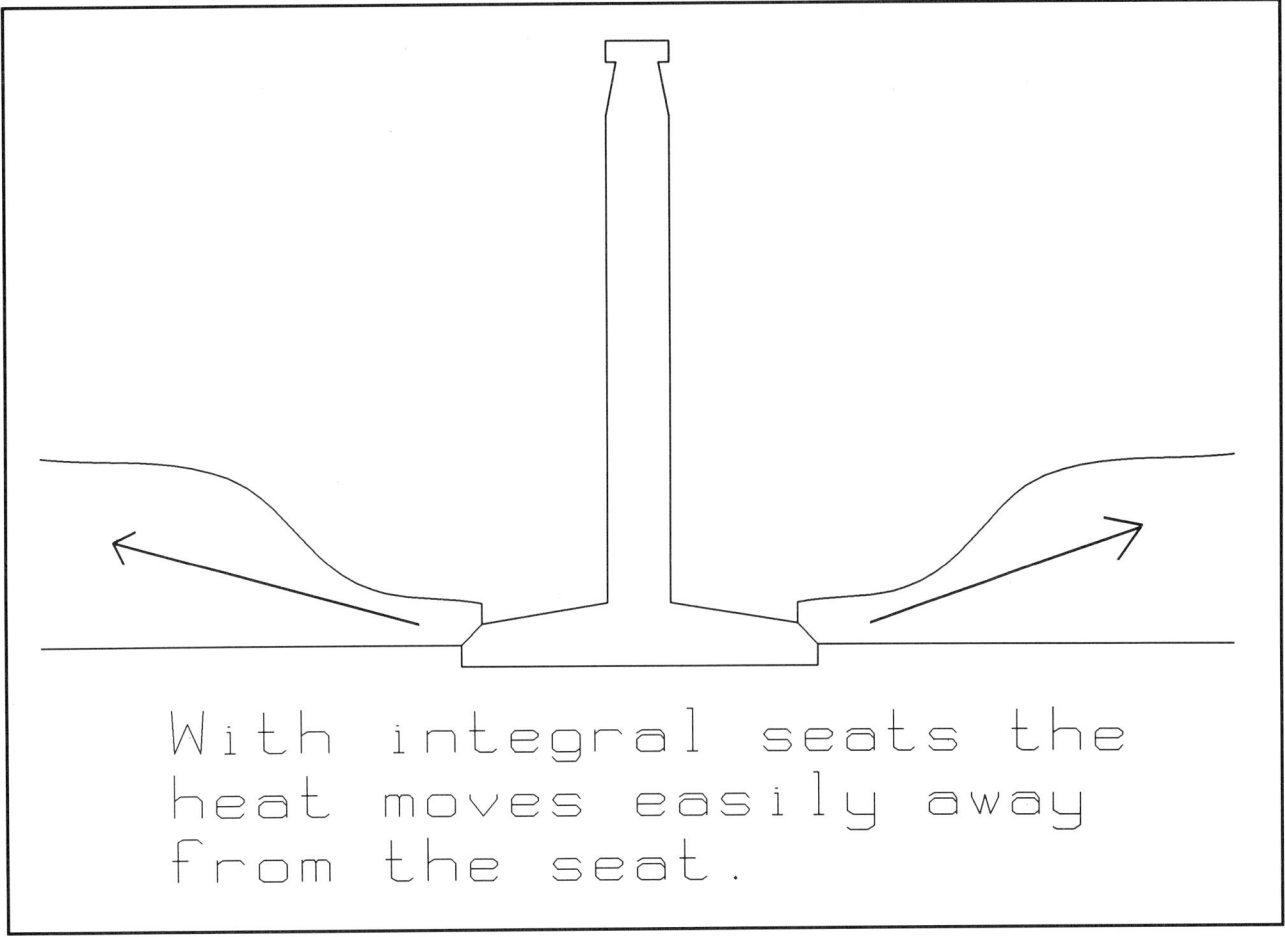

With integral seats the heat moves easily away from the seat.

The integral valve seat has an advantage in its ability to transfer heat from the seat to the head. These seats stay cool.

equal to 5,000-6,000 miles. In a year they run for the equivalent of nearly 70,000 miles. These uses are extreme uses. Long times running with a fixed load and little variation in air flow of engine speed. For these applications alternative valve seats may be advantageous.

Replaceable Seats

Except for a few special purpose applications almost all aluminum or alloy heads use replaceable seats. Some cast iron heads have replaceable seats and some cast iron heads will lend themselves to adaptation to replaceable seats.

There are several problems associated with engineering or re-engineering an application to re-

Many cylinder heads have removable seats. This is especially true of soft alloy heads such as aluminum heads.

placeable seat. At the tender age of fourteen it was demonstrated to me that different types of metal expand at different rates.

Coach Delaney, my ninth grade science teacher, had us measuring the expansion of metal bars. Using a pin and a homemade paper protractor we heated a bar and measured in degrees of the circumference of the pin how much the bar lengthened. Bars of different types had a different rate of expansion. Apply to the previously mentioned VW van. Could it be that the expansion rate of the head and the valve seat were different enough to to allow the seat to loosen in the head?

When replaceable valve seats are used it is because the metal of the head cannot withstand the temperature or the tortuous pounding of the valve. Therefore, by necessity, there will be a difference in the metal of the head and of the seat. If the expansion rates are not similar the head will expand faster than the seat and the seat will fall out of the head. If the seat expands faster than the head it can stress and crack the head.

If the valve seat is too hard it can be brittle and crack from the constant pounding of the valve. The Datsun 280Z engine has a reputation (after many, many miles) for the valve seat being pounded out. When this occurs the valve lash clearance tends to reduce, eventually keeping the valve from fully seating. This can cause the valve and the seat to become overheated resulting in a burned seat and/or valve.

Another reason for replacing valve seats is the use of alternative fuels. Many alternative fuels provide no valve seat lubrication at all. This was a problem when the American public was faced with using unleaded gas in the

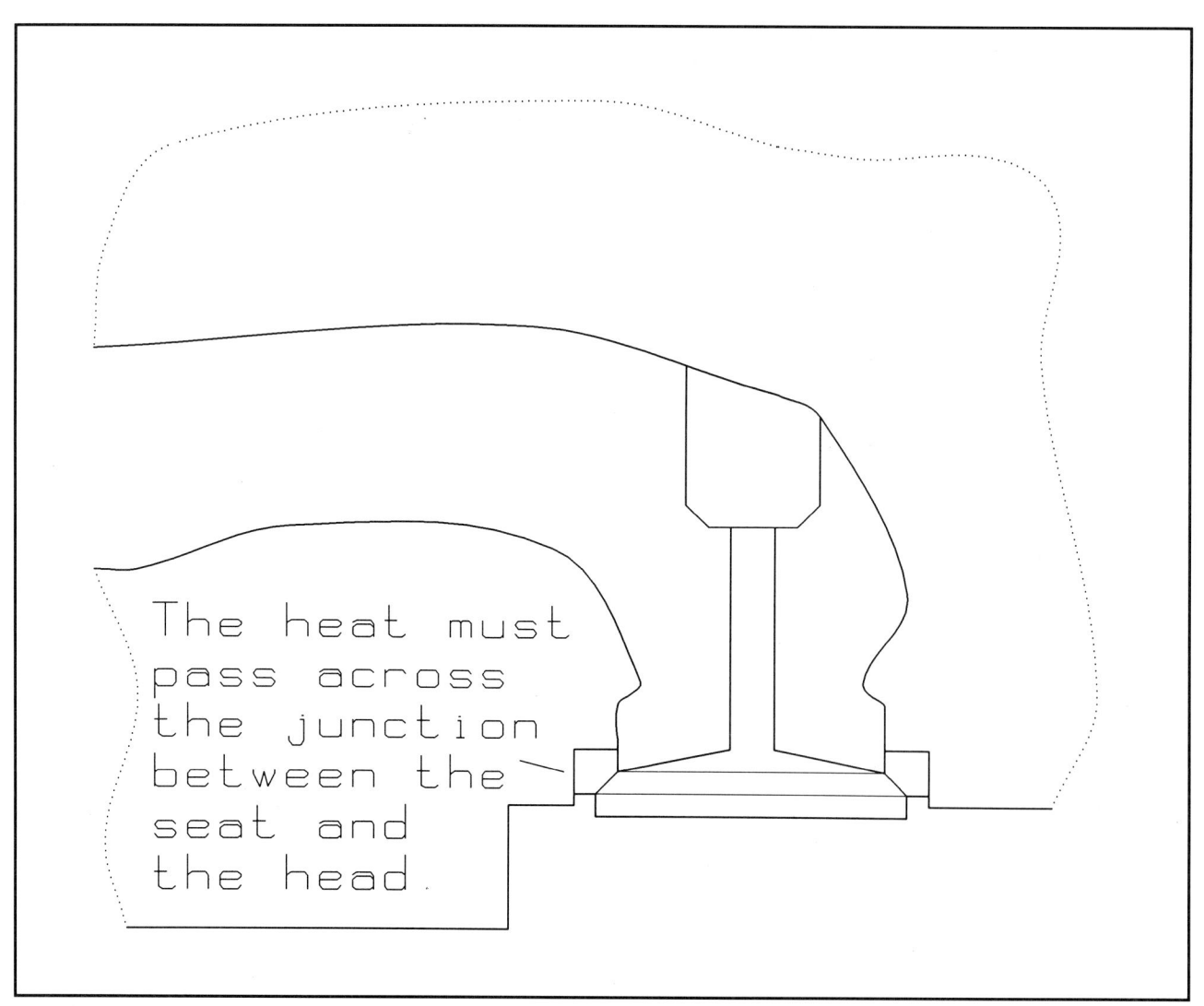

The heat must pass across the junction between the seat and the head.

The removable seat can be easily replaced when necessary, which is handy since it does not dissipate heat as effectively as integral seats.

late eighties. Unleaded gas offered much less protection to the valve seats than did gasoline containing tetra-ethyl lead. For many pre-catalytic convertor engines the reduced valve seat lubrication of unleaded fuel meant reduced engine life. Hardened valve seats were the answer. Alternative fuels such as natural gas and propane have no lubricating characteristics at all. When these fuels are chosen in a fleet environment or by an individual some consideration should be given to the replacement of the valve seats with hardened or sintered seats.

Removable valve seats are sometimes installed in heads with normally non-removeable seats to facilitate the installation of oversize valves. This is often done in heads with removeable seats as well.

When replacing valve seats talk to several machinists and even the manufacturer of the engine if possible. Try to reach a consensus of opinion on what the replacement valve seat should be made of. The use you are planning for the engine will dictate the material the seats should be.

Common materials used in replacement valve seats are:

Bronze

Bronze seats are common as replacement seats on performance engines. On applications where the valve seats are well and effectively lubricated the bronze seats provide a seal and reasonably good durability.

Sintered

According to the *Bosch Automotive Handbook*, second edition, sintering is a process whereby a metal powder is pressed into a desired shape. The shape is then heat treated at 800 degrees to 1,300 degrees Celsius (1472 to 2372 degrees F). This gives the metal extreme mechanical strength. These hardened valve seats are just the thing for applications with high tension valve springs or that use non-lubricating fuels.

Replacing the Valve Seat

Should the valve seat need to be replaced it is time to seek out a higher power than that friend of a friend who knows a guy who does it on the side for twenty bucks.

If the head is designed with replaceable valve seats they must

Replacement of valve seats, especially integral valve seats is not a do-it-at-home kind of job. Large, expensive precision machinery is required.

be removed by cutting the seat with a seat cutter that is one size smaller than the one that would normally be used for this head. As the seat is cut down it will leave a thin wall which can be gently chiseled out. This is a lot like the technique of drilling a hole into a broken bolt that is slightly smaller than the bolt. A chisel is then used to roll the remains of the bolt out of the threads.

101

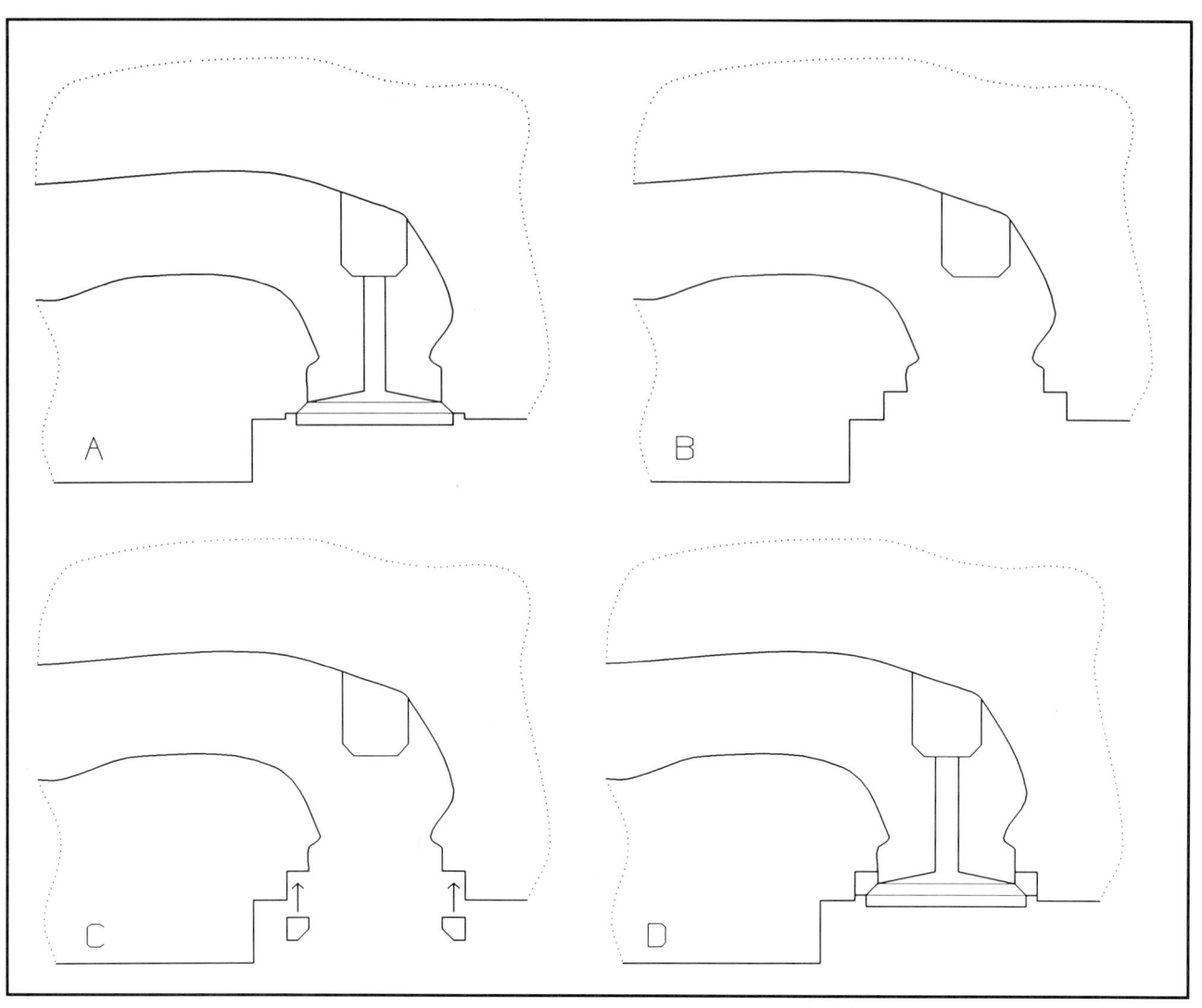

When an integral seat needs to be replaced (A.), the old seat is machined into a platform (B.) for the new seat. The cylinder head is then heated, the new seat is cooled and dropped into the warm head (C.). The new seat now enjoys an interference fit with the old head (D.).

A second way to remove the valve seats is to weld an old, slightly smaller valve to the seat. The stem of the valve can now be used as a drive to drive the seat out.

For those individuals (and your author is certainly not in this category) that are skilled enough to weld two drops of water together a third alternative is possible. Carefully weld a bead around the seat. When the weld cools it will shrink., loosening the seat for removal.

If the head is not designed with removable valve seats, true skill and machine work are required. First a counterbore is machined into the head where the old seat was. This counterbore must be 0.003 to 0.006in smaller than the outside diameter of the replacement seat. The cylinder head must then be stretched around the seat. If your shop is not equipped with a head stretcher, and few on this planet are, place the seat in the freezer overnight. The next day place the head in the oven (yes, my wife would say the same thing) and bake at 350 degrees F for one hour. Remove the head from the oven, remove the seat from the freezer and drop the seat into the head. This technique works great when only one valve seat needs to be replaced. Since the head cools quickly you probably would have to reheat the head several times to replace all the seats with this method. When the head cools and the seat warms up there will be an interference fit that will hold the valve firmly in position. Many machine shops will also peen the head over the valve seat. In reality this does little to hold the valve seat in place, but it makes many of their customers feel more secure.

Grinding the Valve Mating Surface of the Seat

If the valve seats were replaced, and even if they were not, the next step is grinding the mating surface of the seat. The seat on most cylinder heads will be ground at a 45 degree angle. Many machinists and manufacturers recommend that the seats be ground to 44 degrees. Grinding the seat to 44 degrees leaves an interference angle of 1 percent This in theory allows the valve and seat to mate more effectively, mutually machining one another to a good seal within a hundred miles. *However*, when hardened ones are used this angle of interference is not desirable as it may be several thousand miles, if ever, before the valves will seal properly.

At this point grab your tube of Prussian Blue. Place a small blob on the face of the valve, place the valve in the guide. Be sure that you check the valve in the valve seat where it will spend the rest of its life. Press the valve firmly and rapidly into the seat and rotate it with a valve lapping tool. The Prussian Blue stripe that results on the face of the valve should be even and concentric to the margin to face edge of the valve. Additionally, the strips should be centered on the margin leaving an "unblued" band 1/32in wide around the outer edge of the valve margin. The width of the Prussian blue strip sould be between 0.060 (1/16in) and 0.090 (3/32in).

Top cut

If the Prussian Blue strip is too close to the valve margin the seat must be top cut. Use a 30 degree seat grinding stone or tool to lower the contact point of the seat. Repeat the Prussian Blue test as many times as necessary, grind the seat at thirty degrees as many times as necessary, until a 1/32in overhang is achieved.

Chances are that once the 1/32in overhang is achieved the

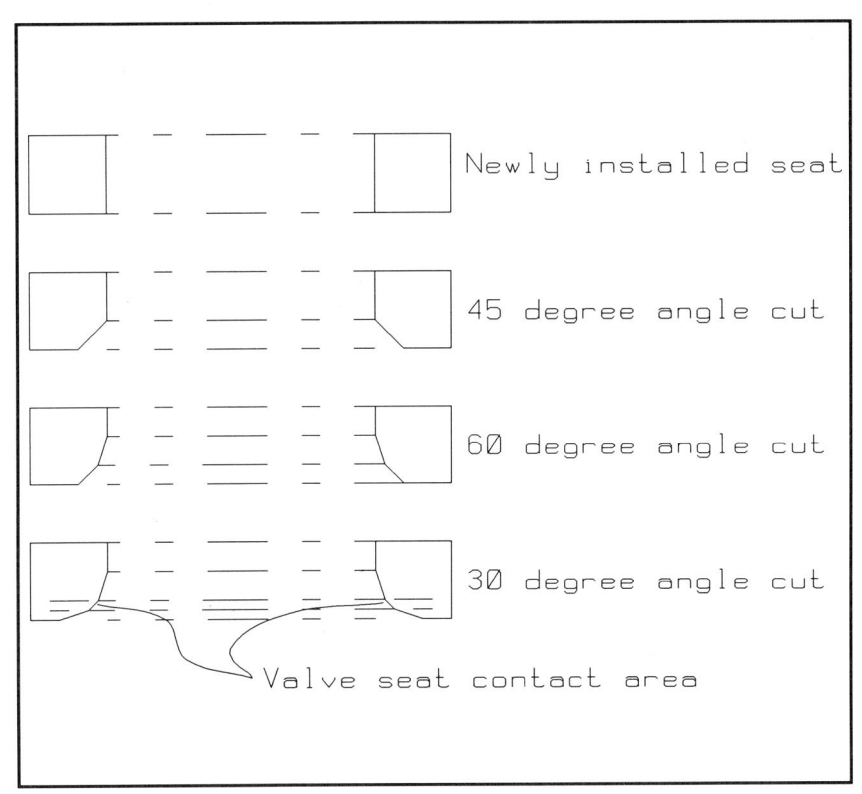

A standard valve grind will leave the seat with three angles. The 45-degree angle will form the contact area with the valve. The 60-degree cut and the 30-degree cut are then used to narrow this contact point and ensure it is in the proper place on the valve face.

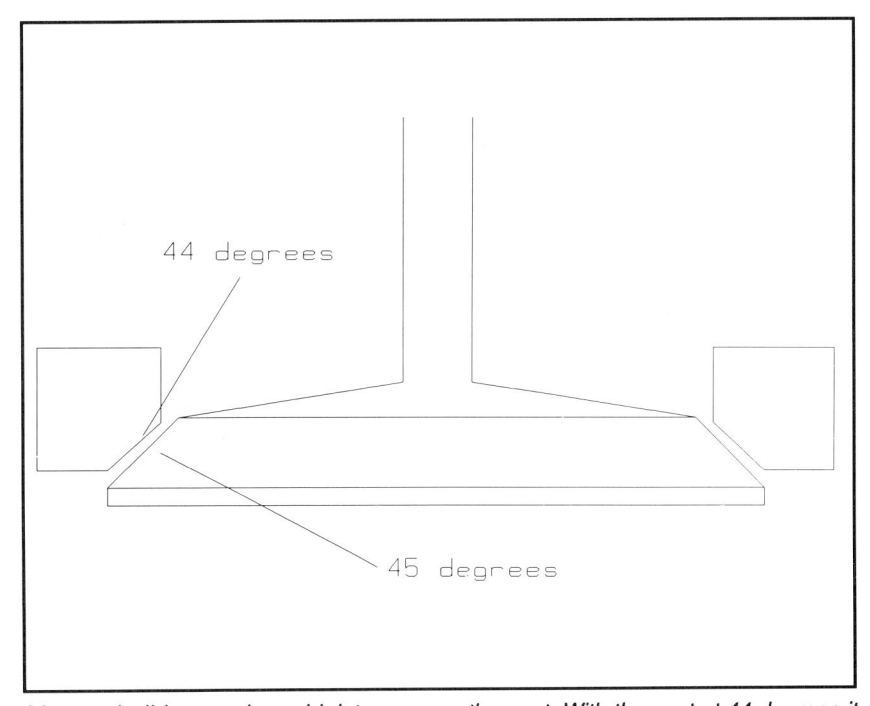

Many rebuilders and machinists recommend a 1-degree interference angle ground between the valve and the seat. With the seat at 44 degrees it is felt that the break-in time for best seal of the valve is reduced.

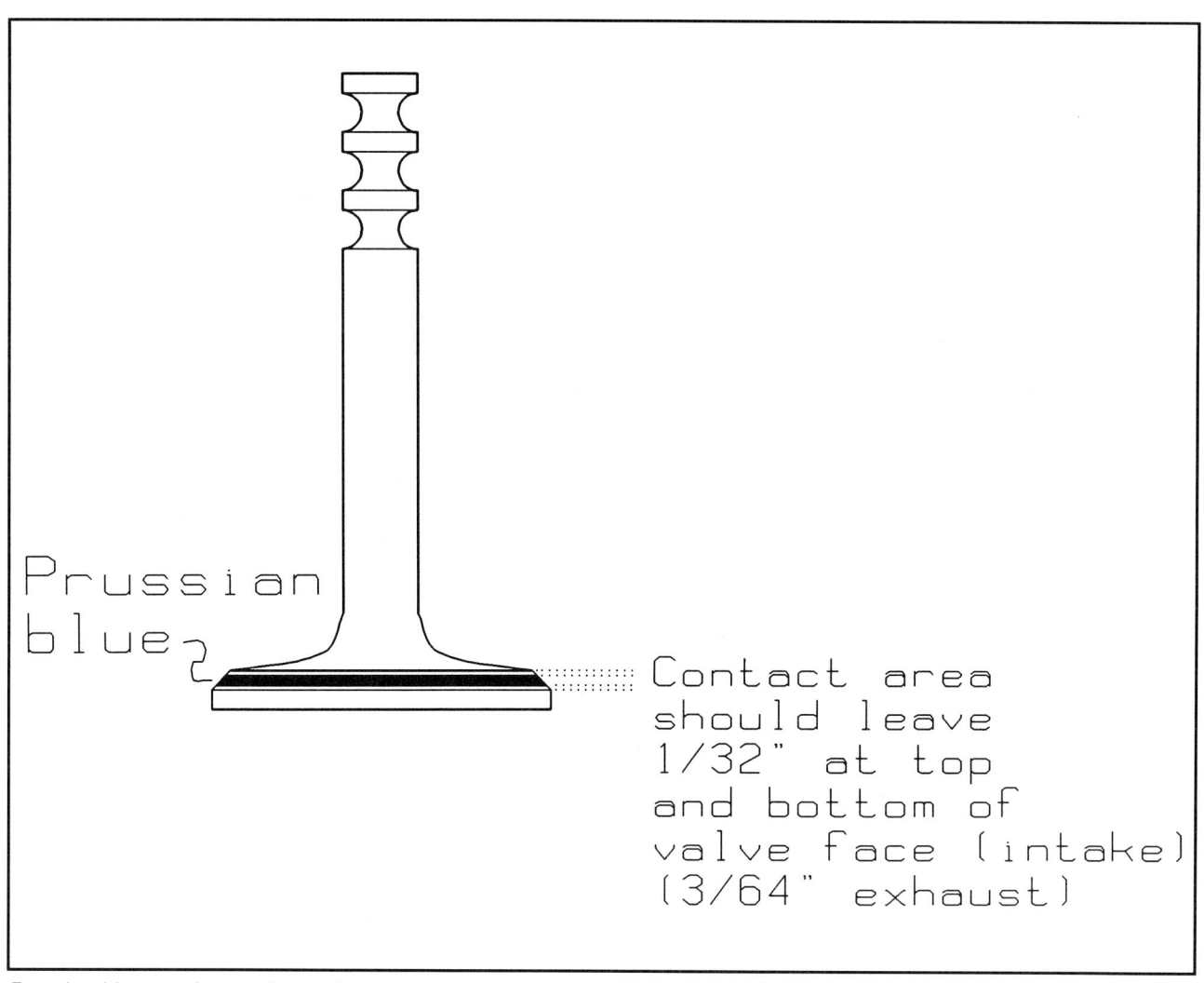

Prussian blue — Contact area should leave 1/32" at top and bottom of valve face (intake) (3/64" exhaust)

Prussian blue can be purchased from most auto parts stores. It is used to ensure the proper contact between the valve and the seat.

seat width will be too narrow. Widen the seat again with the 44 or 45 degree cutting tool or stone.

Bottom Cut

Bottom cutting the valve seat is used to narrow the contact area between the valve and the seat. If the first, or last, Prussian Blue test shows the contact width to be too wide, narrow it using a 60 degree grinding stone or cutting tool.

The operations of grinding the mating surface, top cutting and bottom cutting are what are called a "three angle high performance grind." I have always felt that this is the very minimum that should be done during any valve grind.

Five-angle Valve Grind or Radiused Valve Seat

The truly high performance valve grind produces a "radiused" seat. A radiused valve seat has no sharp edges, instead it features a smooth and round seat that greatly improves air flow. While this is an ideal seat for improving air flow there are very few machine shops that are set up to do a radiused valve grind. If you desire to maximize the air flow yet cannot find a local machine shop that can do a radiused valve grind, a substitute that is usually acceptable is a five angle grind. The standard three-angle valve grind machines surfaces at a 45 degree, a 60 degree and a 30 degree angle. The five angle valve grind machines angles of 45 degrees, 30 degrees, 37 1/2 degrees, 52 1/2 degrees and 60 degrees. The five angle reduces the sharpness of the turn that the air has to make as it passes from the valve pocket into the combustion chamber. While the five angle grind is not as effective at a radiused grind, it comes in a close second. It is also cheaper and very adequate for street performance or the weekend racer application.

Lapping the valves

Earlier I poked a little fun at lapping valves. The process is very time consuming, often heats and expands the valve so that an inaccurate grind is achieved. There is a place for the process after a proper valve grind has been done. Lapping the valves will ensure a better seal on initial start-up. On initial start-up after a three-way grind there may be some valves that do not seat perfectly. This can cause a rough idle after start-up. After a few minutes of running heads using standard valve seats the valves will seat properly. If hardened seats have been used, it will take as much as several hours for the valves to seat properly. If the engine is to be used in racing, the total life expectancy of the engine may be measured in hours. There may not be time for the valves to seat. If the engine has been rebuilt for a customer, or worse yet, a neighbor, the valves never will seat. Although the engine smoothes out by the time he brings it back for a 500-1,000 mile check-up, what will stick in his mind is how it ran when he first drove it. Nothing that happens after that will ever convince him that the engine is now running properly.

To lap the valves use a valve lapping stick and valve lapping compound. A lapping stick is a wooden or plastic stick with a suction cup on the end. Frankly, it looks like a ridiculous piece of equipment for a professional auto repair shop to have in public view. Valve lapping compound is liquid, almost pasty substance containing a highly abrasive grit. Put a small daub of lapping compound on the valve seat, dampen the suction cup on the end of the lapping stick and press the lapping stick onto the end of the valve. At this point flash back to kindergarten, remember how you made snakes with modelling clay? Roll the lapping stick back and forth between your hands. This will polish the mating surface between the valve and the seat. For those who skipped kindergarten, and therefore missed out on learning one of life's essential skills there is an alternative.

Slip a piece of 5/16in (or 8mm) fuel line over an old drill bit. Chuck the bit into a drill motor then slow the other end of the hose over the valve stem as it protrudes through the head. Run the drill on the lowest speed while pressing the valve firmly into its seat with a dowel.

Runout

Runout is a measurement of eccentricity. Precision machined rod with a dial indicator is placed in the valve guide. The dial indicator then runs around the 45 degrees (or 44 degrees) surface of the valve seat. This checks two things, the concentricity of the seat and confirms the guide and seat are at perfect 90 degree angles. Now, if all of the above operations, replacement of the valve seats and the machining of the valve to seat mating surface have been done properly, both of these things should be true. So why check the run-out? Simply to confirm that all of the above *are* true.

Valve depth

Remember that we are attempting perfection in this perfor-

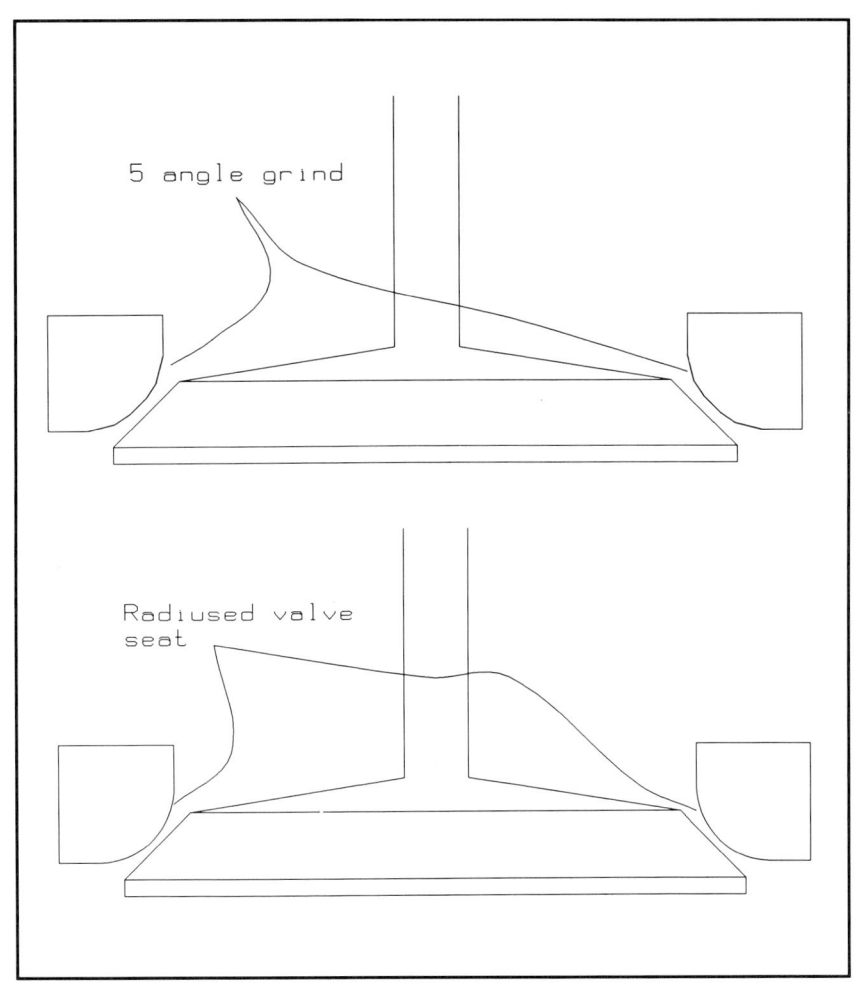

A true performance valve grind removes the sharp angles inherent to the three-angle grind. Two alternatives are the radiused seat, which few shops are equipped to do, and the five-angle grind, which virtually any shop can perform.

105

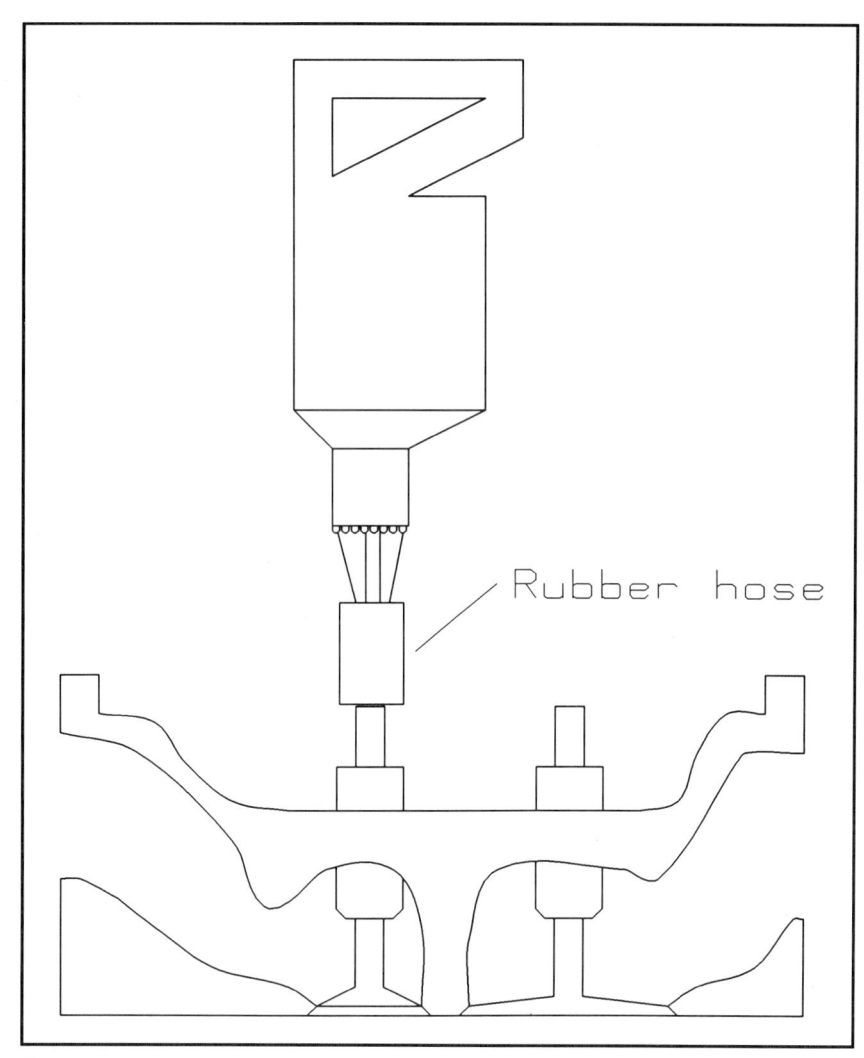

Slip a piece of 5/16in (or 8mm) fuel line over an old drill bit. Chuck the bit into a drill motor, then slip the other end of the hose over the valve stem as it protrudes through the head. Run the drill on the lowest speed while pressing the valve firmly into its seat with a dowel.

mance valve grind. Perfection means that the volume of the combustion chambers in the heads should all be the same. The part of the valve below the lower point of contact between the valve and the valve seat is in the combustion chamber and will affect its volume. To ensure that each of the valves protrudes the same depth into the combustion chamber use a dial indicator mounted on a bar that will fit across the combustion chamber. Each of the intake valves should have the same depth. Each of the exhaust valves should have the same depth. I know what you are thinking, "How nit-picky can you be?" In all seriousness, the intent of blueprinting a head, or in other words doing a true performance valve grind is to ensure that every possible bit of horsepower is squeezed from the engine. If there is a discrepancy between the volume of the combustion chambers there will be less power from the cylinder with the smaller volume. Additionally there will be an inherent roughness to the operation of the engine.

If the valve depth is imbalanced, the seat on the highest valves will need to be reground or recut to provide the proper height. Grind the valve seat at a 45 degree angle. As you are grinding the seat, check the valve depth frequently. When the proper valve depth is achieved, top cut and bottom cut the seat as described earlier to provide the proper valve to seat contact.

Valve Masking

When planning this book it was difficult to determine the proper place to talk abut valve masking. Valve masking and the reduction of valve masking are among those operations that should be done as a part of a port and polish job.

Due to the basic design of the overhead valve engine there will always be some valve masking. Valve masking occurs when the proximity of the valve to the side of the in-head portion of the combustion chamber. Valve masking is the effect of a pressure that builds up between the valve margin and the walls of the in-head portion of the combustion chamber.

The amount of air flowing through the valves in and out of the cylinder head should only be limited by the valve lift. If the valve lift is greater than the clearance between the margin of the valve and the side of the combustion chamber, then this clearance and not valve lift will limit air flow through that side of the valve.

Most diesel engines offer a virtually perfect situation when it comes to valve masking. These cylinder heads are perfectly flat on the bottom, there is no part of the combustion chamber in the cylinder head. The result is that there is no part of the cylinder head combustion chamber that interferes with the flow of air and exhaust gasses around the valves. However, even in these engines masking can occur. The valve hangs into the cylinder when it is open and the cylinder

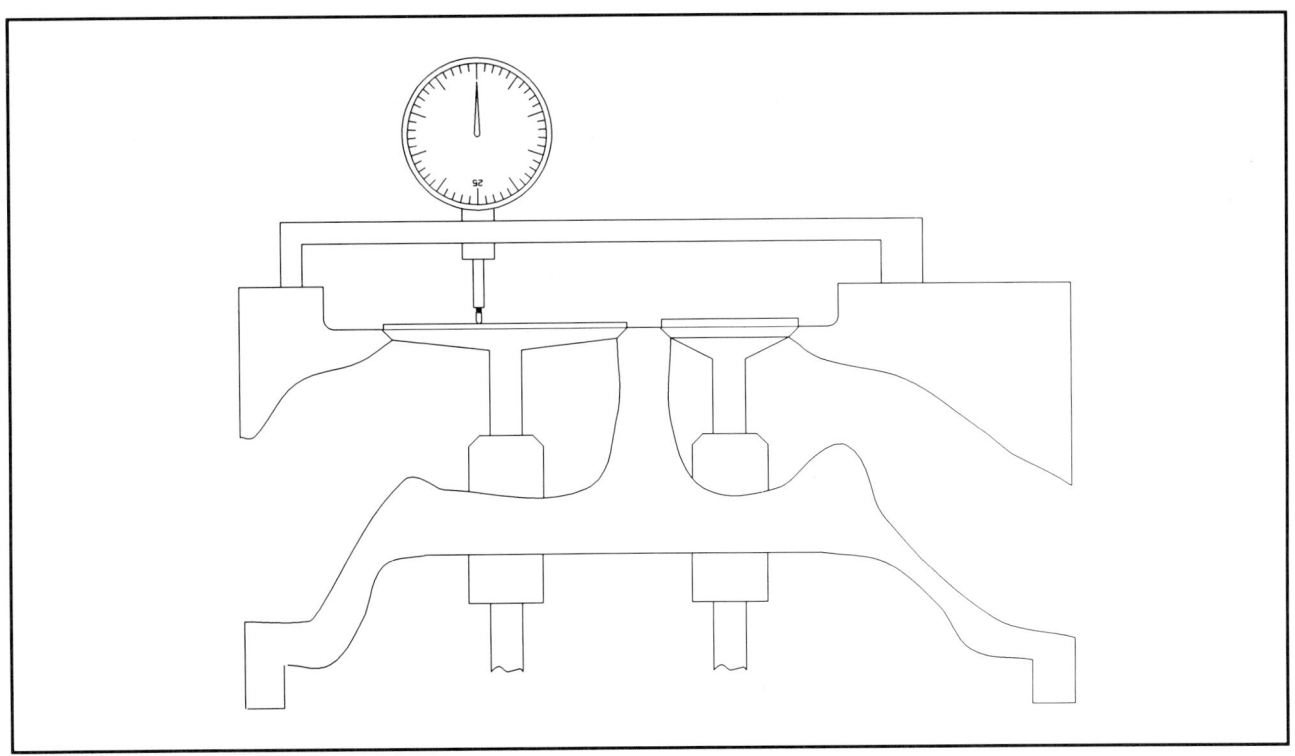

Valve depth can affect the size of the combustion chamber. If you are truly trying to reduce the stress on the engine and maximize performance, all of the intake valves should protrude the same amount into the combustion chambers and all of the exhaust valves should protrude the same amount into the combustion chambers.

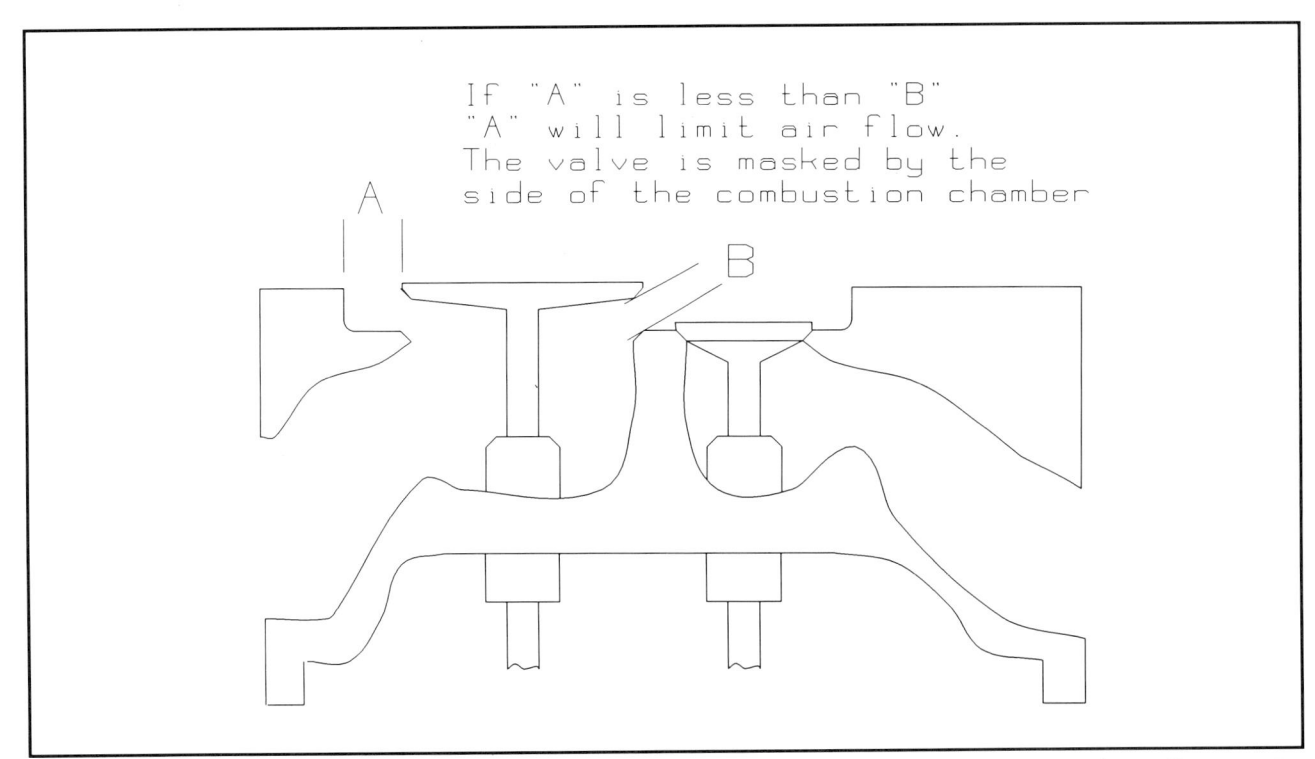

Valve masking is when the distance between the margin of the valve and any other object, such as the side of the combustion chamber, is less than net valve lift. When this happens, that object, rather than valve lift, will determine how much air can flow past that portion of the valve.

107

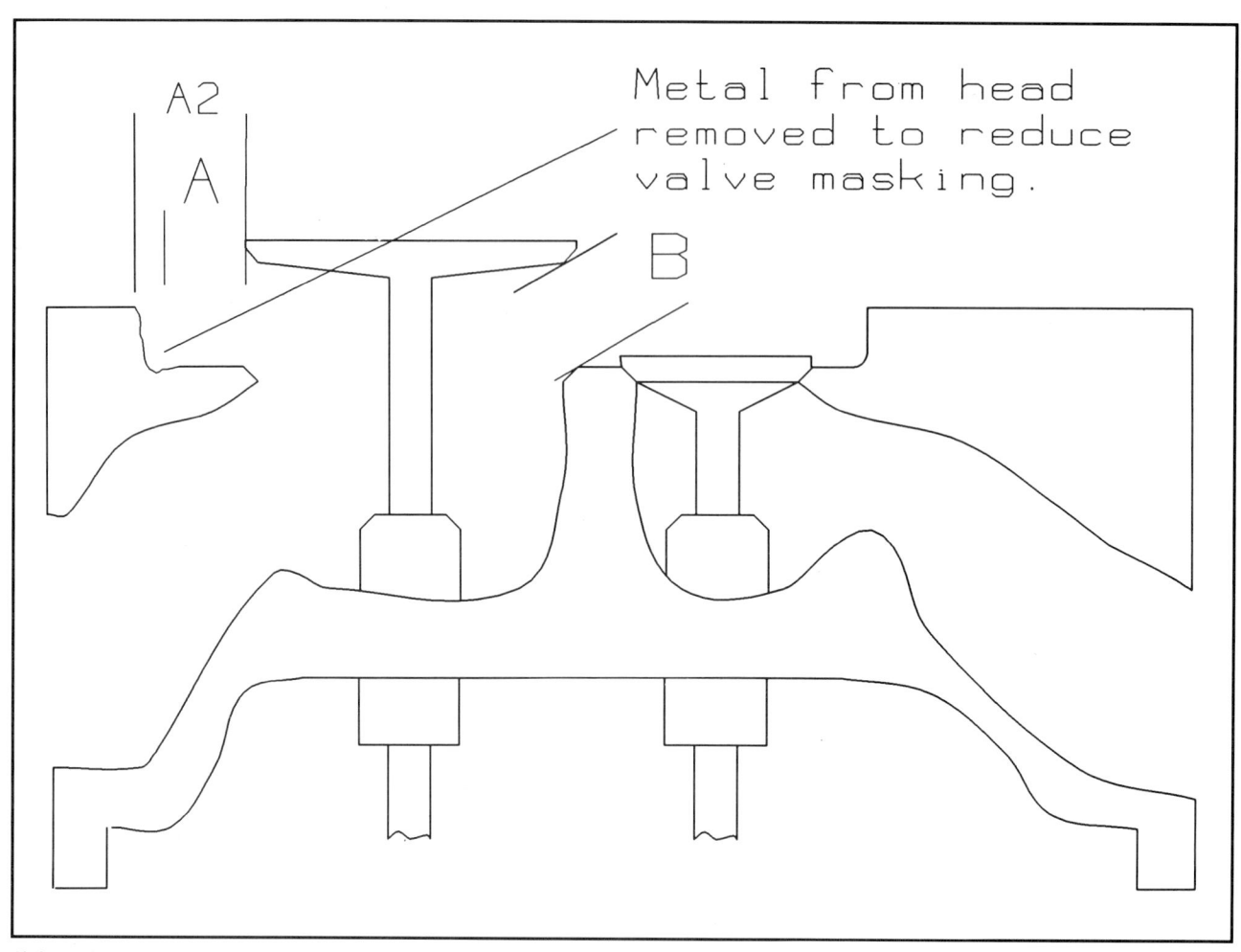

If there is no danger of penetrating the head into a water jacket or oil gallery, enough metal should be removed from the sides of the combustion chamber to unmask the valve. One BIG caveat: Will the benefits of unmasking the valves be less than the negative affect of the removal of the metal lowering the compression ratio or causing an imbalance in combustion chamber volumes?

When grinding metal from a head, either to alter the intake or exhaust porting or to alter the combustion chamber size or shape, use a high-horsepower, high-speed grinder. That 1/4-horse electric drill that the kids gave you last Father's Day probably will not work.

wall itself can mask air flow around the valve. This type of masking problem can only be resolved by redesigning the heads and relocating the valves. At least on most gasoline engines the problem can be reduced or maybe eliminated.

Resolving the problem of valve masking is very expensive or time consuming. This is the first major operation of working a cylinder head that really separates the mature articulate primates from their off-spring, or as the politically non-correct might say, the men from the boys. First you need to acquire the proper power tool. I have seen the re-

108

sults of guys trying to port, polish, or otherwise re-work their cylinder head using the quarter-inch drill motor that their wife bought them for Valentine's Day to remove the last excuse for not hanging the portrait of Uncle Ed in the living room. It is not pretty. This operation requires a high-speed electric or air powered tool with power rating in horsepower not fractions. The grinding tool itself should be a high speed carbide grinding bit.

In the ideal world you would now obtain an x-ray of the areas of the head immediately around the combustion chambers. I have often wondered how my doctor would react if I asked him for a radiologist referral to X-ray a pair of Chevy heads. Second best would be accurate drawings of the inside of the head locating water jackets, oil galleries and other hazards. You might as well try to find an efficiently run government program. Without these things you are at a disadvantage, you are at the risk of grinding through the metal into one of these areas. Trust me, you will be at a disadvantage, you will be at risk.

Paint the bottom of the head with a light mist of your favorite color of paint. There are professional dyes designed specifically for this purpose. Carefully measure and scribe the outer limits of the metal you are going to remove to resolve the masking problem.

As you begin to remove metal remember it is important that the size of the combustion chambers be of equal volume. This is more important than eliminating the masking.

Watson's corollary to Stephen Hawking's musings on alternate miniature universes: Alternate universes exist within every cylinder head. As metal is removed from a cylinder head these universes migrate into the head manifesting themselves as holes between the combustion chamber and the water jacket.

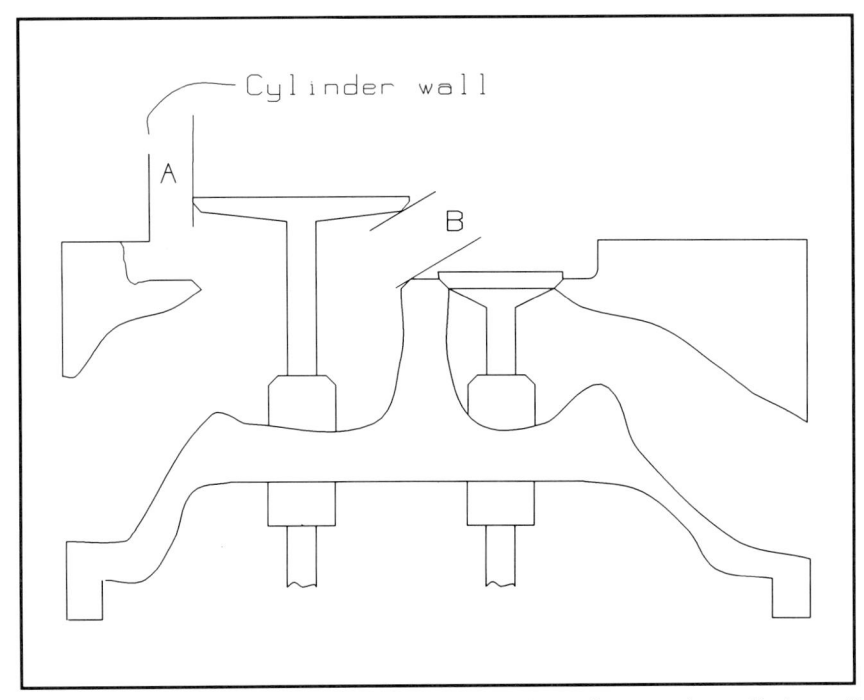

Of course if you are using a high-lift camshaft and spend a great deal of time redesigning and size-matching the combustion chambers, you may do so only to discover that cylinder wall now masks the valve and all the head work was for naught.

To avoid being affected by this obscure law of the universe talk to as many machinists or people who have performed this operation before as possible.

Grind the area that is masking the valve. Try to get a clearance between the valve at all points of the opening process that is equal to the valve lift. Calculating the valve lift will require knowing the cam lift and rocker ratio. Therefore the cylinder head work again should not be done until the cam and valve lift decisions have been made.

I hope that you do not have 4lb of aluminum shavings on the floor before you read this paragraph. Every ounce of metal you remove from the combustion chamber reduces your compression ratio in that cylinder. Are you frustrated yet? Again, each time you resolve one problem you affect another aspect of the performance modification. As you remove the metal, CC the cylinder head combustion chambers several times. Each time you CC the head calculate the new compression ratio. Keep in mind also that the compression ratios must be equal in all cylinders to ensure smooth performance. Do you need a Valium yet?

To calculate compression ratio:

Compression Ratio = (Cylinder Volume + Chamber Volume) / Chamber Volume

Displacement Ratio = Cylinder Volume / Chamber Volume

Summary: While valve masking can limit air flow through the valves there are plenty of other considerations that may influence your decision to resolve the problem. Will the loss of compression be made up for by improved air flow? Are you skilled enough to keep each of the combustion chambers equal in volume so the compression ratios will all be equal? If you have an electric grinder and do not have cable will your spouse put up with hours of snow on the TV screen while you work your performance miracle?

Porting and Polishing

This is a phrase that I first read in a "Dennis the Menace" comic book in 1961. In this story Dennis's dad had taken him to the go-kart track. One of Mr. Mitchell's friends was telling him what he had done to the go-kart engine. Among the things he mentioned was porting and polishing the head. I knew at that moment it was my destiny to port and polish something. In less than twelve years I had realized this goal, actually my racing partner did the porting and polishing. In 1973 I really did not have the patience for such things.

This term "porting and polishing" is often thrown around very haphazardly. The term is used for every everything from an intake/exhaust spit shine to a full blown flow-benched, meticulous job.

If you have a cylinder head beside your bed as you are reading this book, look at it. If not, wander out into the garage and look at the cylinder head sitting on the workbench. Notice that there are parts of the intake and exhaust ports that are very to get to, there are other places that are very difficult. It is those difficult places that require most of your attention.

The Proper Tools

Again a high speed grinder and a carbide grinding bit are needed for porting the head. Do not expect to find such a tip on the bargain table at the local grocery. These are quite expensive, but quite durable. You will need a second, finer bit for polishing the ports.

Porting

This operation involves the reshaping of the cylinder head intake and exhaust ports to improve air and exhaust gas flow. When the head was designed by the manufacturer a great deal of attention was given to port design and air flow. Then, depending on the manufacturer a degree of compromise was given to requirements for mass production. As a result there are many imperfections in the design and execution of each head. The goal of porting the head is to ensure that the potential flow rate through

Porting and polishing are the black belt operations of cylinder head martial arts. The maximum gas flow rate through this stock exhaust port could be greatly improved with a little careful porting and polishing.

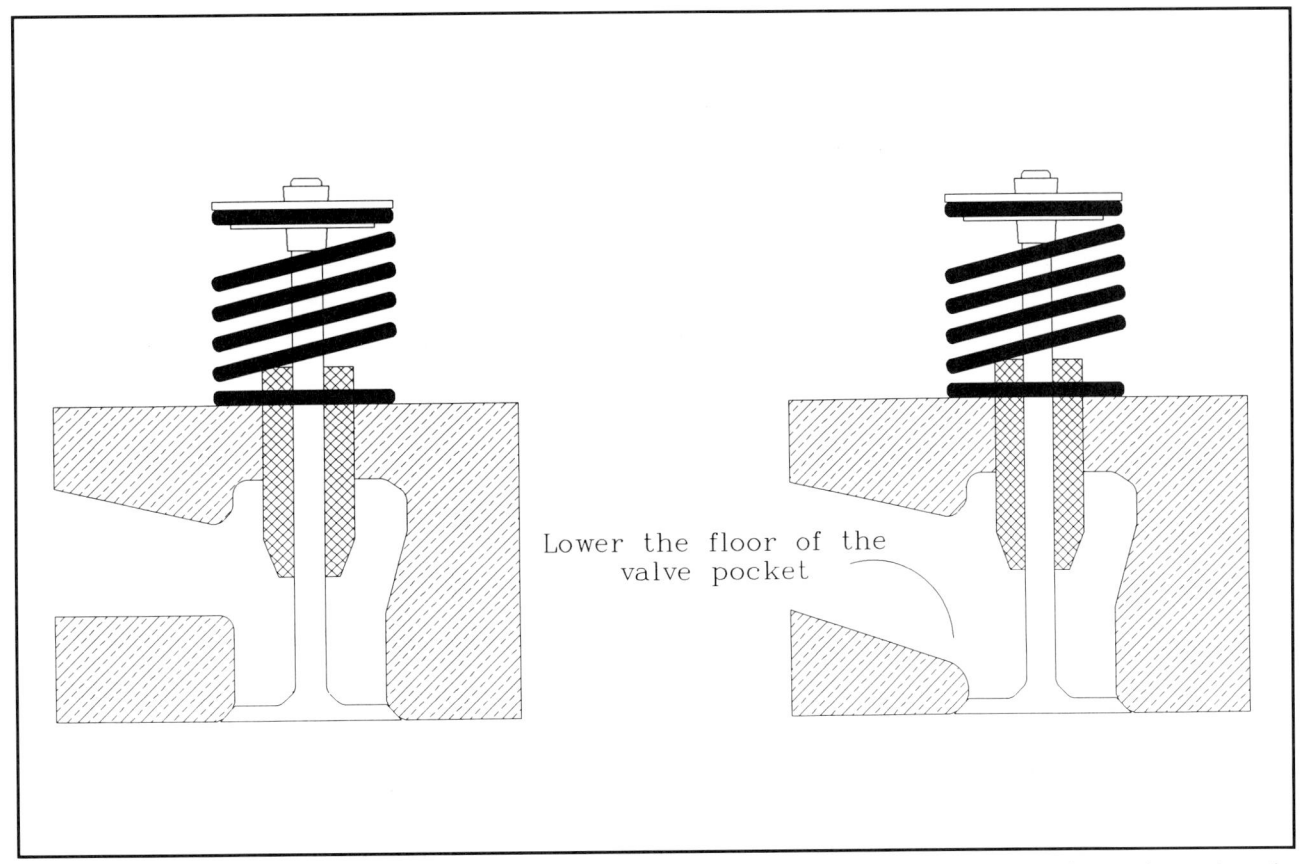

Eliminating the sharp turns that have to be taken by the gases flowing in and out of the combustion chamber is the most important part of porting the head. In the illustration the valve pocket floor, known as the short side radius, has been lowered to reduce the angle of the turn.

the intake and exhaust ports is greater than the theoretical flow rate past the open valve at maximum valve lift.

The air flow rate when the valve is at maximum valve opening is represented by the surface area of the imaginary cylinder formed between the top edge of the valve margin and the bottom edge of the valve seat. If the maximum valve lift is 0.5in and the valve diameter is 2in, then a cylinder 0.5in tall and 2in in diameter is formed. The formula to convert these measurements to a surface area is:

π x Diameter x Lift

or for this example:
3.14 x 2 x 0.5 = 3.14 square inches

The significance of this is that if we have a straight path for the air to follow, and if the width of the intake port is 1.25in at its narrowest point, the height must at no point be less than 2.09in.

So: Open Valve Area / Width = Minimum height

Any point that is smaller will create a restriction greater than the valve. This formula assumes a straight flow of the air through the head. There are very few heads, other than those designed specifically for racing engines, that have straight ports.

The next consideration in porting the head is to compensate for turns, bends, twists and obstructions to the air flow. In reality these are the hardest to deal with when porting the heads. Rethinking the compromises made by the engineers can cause problems on some heads. There are

This intake port of a Winston Cup head shows the scribe marks used to size the ports to the intake manifold and the intake manifold gasket.

111

A virtual work of art, this Winston Cup head features enormous valves and exhaust ports half the diameter of the Hampton Roads tunnels.

heads where there is a rise in the floor of the port. This is an obvious place to lower to improve air flow. Be cautious, carefully try to assess whether or not the rise is there to accommodate an oil gallery or water jacket. There really is no accurate way to predict the effect of these nonconformities, you must eliminate as much of these problems as possible and flow test the head to ensure that the desired results have been achieved. Be prepared to have the head flow benched three or four times before the desired results are achieved.

The Congressional money saving technique number 1: Tell everybody (wife, husband, parent, customer) you are planning to have the head flow tested four times at a cost of $98 each time. When the head only needs to be tested twice you have *saved* them (wife, husband, parent, customer) $196. It works for Congress and it will work for you.

You can save even more money by porting the first intake and exhaust port, having them flow-benched to ensure the proper air flow. Now match the shape of the other ports to these.

In porting a head for street or track performance use there are three areas that need attention. These are the short side radius, valve bowl and the intake ports. Use a carbide grinder to smooth the area immediately behind the valve, as the seat transitions into the intake port and around the bottom of the valve guide. When the head is manufactured this area is left rough. The roughness disturbs the air flow into the cylinder. Smooth this area while trying to remove as little metal as possible.

The Short Side Radius

The short side radius is the bump in the floor of the intake port of the head. It is the area where the air has to turn the corner on its journey to the cylinder. Lower this hump, make the corner that has to be turned as shallow an angle as possible. In fact, you are trying to straighten out the turn to improve air flow. At high velocities the air responds to this hump in much the same way as the air would respond to a wall in the port. If the short side radius rises 0.5in into the straight-line shape of the port, and if the height of the port is 2.0in, then the port height is effectively reduced by one fourth. In reality the effect is not really that great because it is unlikely the air flow rate through that section of the port would ever be at the maximum for the port. Calculate your air flow requirement based on displacement and maximum desired rpm and ensure when the head is flow tested that its flow rate exceeds requirements of each cylinder.

While it is not always a good idea to match the size of the intake port to the intake gasket, doing so does illustrate how much this intake port could be improved.

[(Displacement / 1,728) x RPM] / 2 = CFM

CFM x 1.5 = Requires Head Airflow

(*Note:* displacement is in cubic inches.)

The Valve Bowl

The valve bowl is the area behind the valve where the valve guide protrudes into the port. The valve guide is positioned in a boss, or mount, that protrudes into the air stream on most cylinder heads. Additionally the valve stem itself inhibits air flow. Early in this book I described the ideal head. It would be a force field which would be switched off during the intake and exhaust strokes. The second best head

Port and polish the first intake and exhaust port, then before doing the rest, have this cylinder flow tested. If the results meet your needs use these first ports as a model for the rest. Since most places charge for flow bench time by the cylinder or by the hour, this can save a lot of money.

113

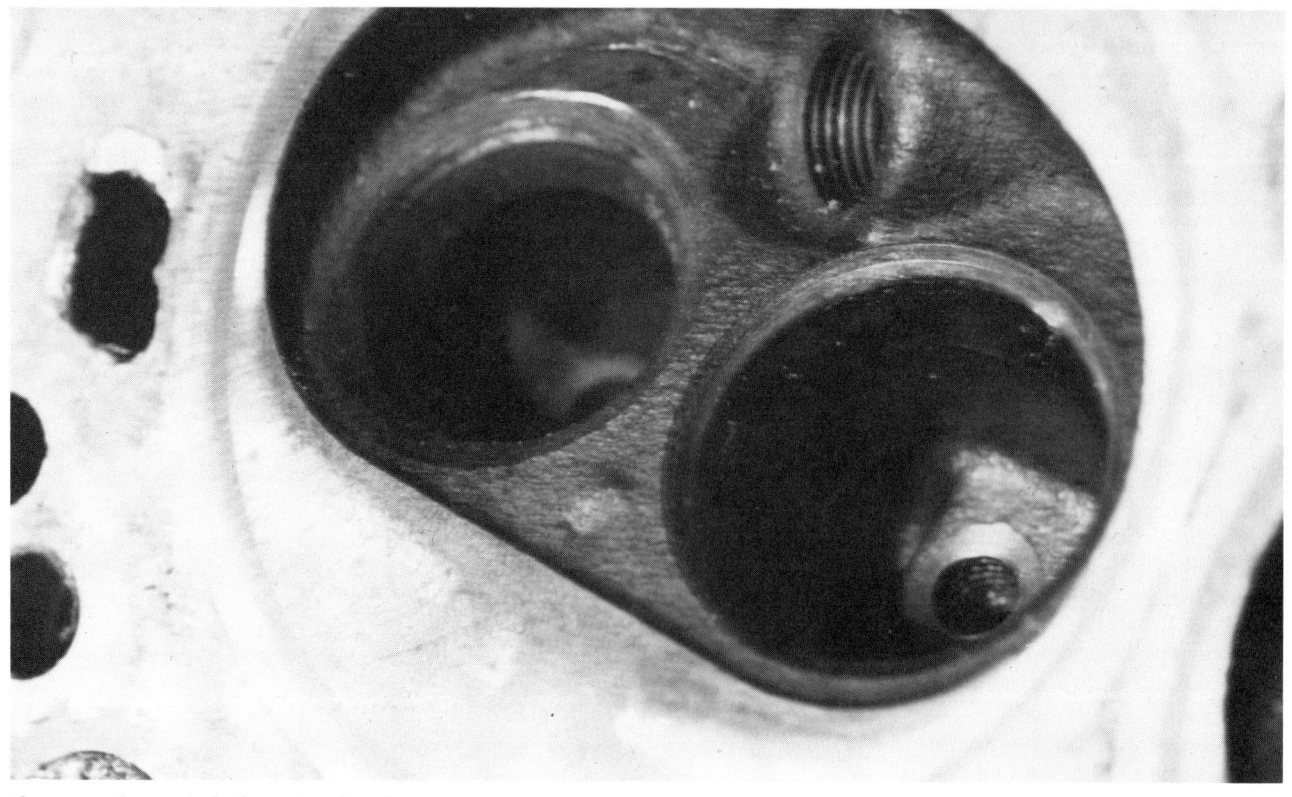

Compare the ports in these two heads. One is clean and smooth. The other needs a great deal of work.

would be one without a guide or valve stem. Unfortunately this is not possible. Therefore, we must minimize the valve guide boss and leave the valve stem. Taper and streamline the valve guide boss.

Resizing the Ports

Finally use the carbide grinder to open the intake port to match the intake manifold gasket. Paint the bottom of the head with a light mist of your favorite color of paint. There are professional dyes designed specifically for this purpose. Carefully measure and scribe the outer limits of the metal you are going to remove to match the head to the intake and exhaust manifolds.

If you need to enlarge the port, and there are many heads which will not require port enlargement, emphasize enlargement of the top; make the top of the port higher and wider. Transition this enlargement smoothly into the valve bowl. If the bottom of the exhaust port does not match the exhaust manifold gasket, do not try to make them match. The misalignment normally found here is intentional, it forms a reversion dam which keeps the exhaust gasses from reentering the combustion chamber as the exhaust pressure pulses vary.

Cardboard Templates

Once the desired shape for the first intake and exhaust port has been determined, cut and flow benched, you will want to match the characteristics of the other ports to this port. From stiff cardboard or plastic, make a set of templates for the short side radius, the manifold port width at the top, the manifold port width at the bottom and the height of the manifold port. Additional templates should be made of the valve pocket and the guide boss. The real art and what will really prove your skill will be how closely you can match the other ports to this port.

Checking with Silicone Goop

By now you finished grinding and trying to match the ports with the cardboard templates. If you are anal retentive you have probably spent several weeks perfecting the ports, if you are a ritalin candidate you have probably changed your hobby from performance engines to quilt making. If you are an anal retentive ritalin

This head is ready for competition. It even looks polished doesn't it?

Burette

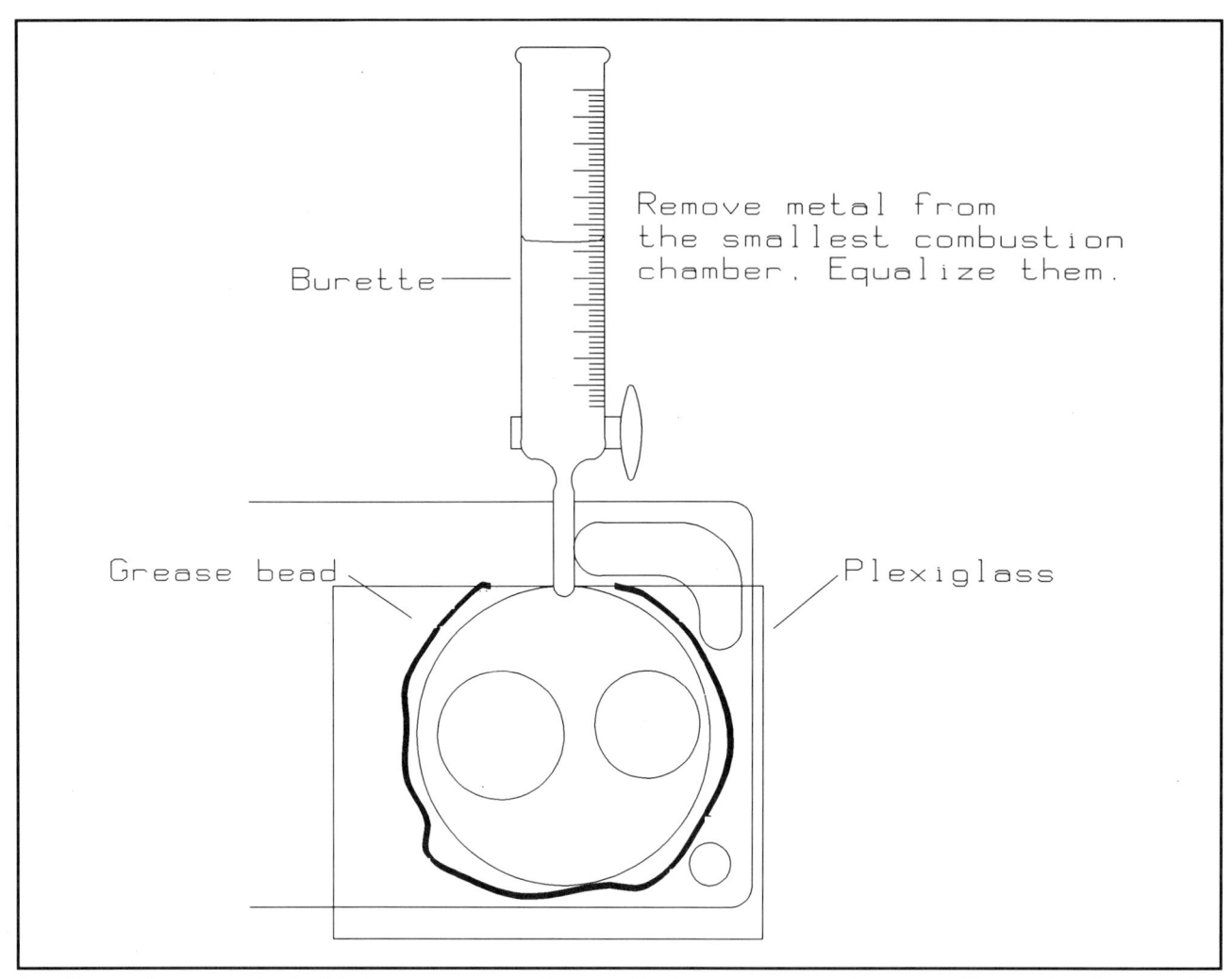

Remove metal from the smallest combustion chamber. Equalize them.

Grease bead

Plexiglass

I really consider any engine built today to be a performance engine. If the heads you are building are destined for the top of a 454 to be used at the Spring Nationals, the need for precision is obvious. But today performance can mean improved fuel economy or improved emissions. No matter how you are defining performance, the size of the combustion chambers will be critical to how much power each cylinder produces. If one of the cylinders has a smaller combustion chamber than the rest it is a burden on the other cylinders and as such can affect power, fuel economy, and emissions.

candidate (like your author) you are doubtless talking to yourself. This is one skill where dedication, patience and the desire for perfection are absolutely necessary.

Now it is time to check your work. Chicago Latex and Permaflex Mould Company both make a latex compound that is used to check the relative size and shape of the ports. Spray all the surfaces of the intake and exhaust ports with WD-40 or similar lubricant. Slant the head lengthwise slightly so that one end of the head is slightly above the other end of the head. Tilting the head reduces the possibility of bubbles being trapped in the latex when it is poured into the port. Place the valves in their appropriate guide. Suspend the valves so they are high enough above the seat to permit pouring the latex into the valve pocket and port. Place duct tape over the intake and exhaust ports. Mix the latex according to instructions provided with the latex. Fill the intake port with the compound through the open valve. Pour slowly to reduce the possibility of an air bubble being trapped. Once the intake port is filled, fill the exhaust port. Once both ports are filled, fill the combustion chamber portion of the head. Now push both valves through the latex firmly into their seated positions and finish filling the head combustion chamber. Repeat this process for the other cylinders.

Once the latex is set carefully remove the molds along with the valves. When they are all out you will have a 3-D representation of the ports. Now retouch the ports to match your first "model" port.

Polishing

When I hear the word "polishing" I think of green and khaki, rude men with gold stripes covering their arms from shoulder to wrist, brass and jump boots. "Son, if I can't see my face in those boots tomorrow you are running to the mess hall." Why would he want to see *that* face and when did the mess hall ever have anything worth running to?

The high luster "spit shine" that these images invoke are not what is referred to when we mention polishing the ports. In fact, it have been shown in many industries that when a surface is too polished it can impede rather than help fluid flow. Also if the surface is slightly rough it creates turbulence that helps keep the fuel suspended in the flowing air. When you polish the head you are looking for casting ridges and places where the walls of the port become abruptly rougher than the rest of the port. Ridges and sudden increases in the roughness of the surface can decrease air flow through the port.

Using the grinder motor that was used to port the head and a grinding bit that is finer than the one used to port the head remove the casting ridges and insure that the steps and grooves that might have while porting the head are smooth to allow good flow.

CC the Combustion Chambers

The next step is to ensure that each of the in-head portions of the combustion chambers are equal in volume. This process was briefly mentioned earlier in the book. This is the point where the detail work should be done. Get a CC gauge. A CC gauge is known to chemists as a graduated burette. A burette is a glass tube marked in milliliters. At the bottom of the tube there is a petcock. When the burette is filled with mineral oil and slowly drained through petcock into the combustion chambers of the cylinder head it is possible to measure the volume of the combustion chambers. When blueprinting an engine, the equality of combustion chamber volume is critical to achieving equal power from each of the cylinders.

Place a piece of Plexiglass over to combustion chamber. A little petroleum jelly will form a very nice seal between the head and the Plexiglass. Place the top edge of the Plexiglass just below the top edge of the combustion chamber. Slowly fill the combustion chamber with mineral oil. As you fill carefully monitor the amount it takes to fill the combustion chamber. Note these amounts for each of the combustion chambers. Do not change the size of the largest chamber, the one that takes the largest amount. Enlarge the smaller combustion chambers. A logical place to grind away the metal from the combustion chambers is from the area masking the valves.

Summary of Porting

I teach skills upgrading classes to journeyman automotive service technicians in subjects ranging engine rebuilding to electronic fuel injection. One of the more common comments I get from the students in these classes is, "Gee, you sure make that sound easy to do, but I'll bet it isn't really that easy."

Watson's Theorem of Real Life, Corollary 16: The easier it sounds in class, the easier it sounds on paper, the easier it sounds in class, the longer it will take to do it.

Major Valve Refitting

12

Oversize Valves

At first, the idea of putting in oversize valves may be quite appealing. Larger valves should guarantee greater airflow, right? Not necessarily. This holds true only if valve masking with the larger valve can be kept at a minimum. It doesn't help to increase the diameter of the valve by 10 percent if the larger valve ends up too close to the combustion chamber wall. Will larger valves increase airflow, or will the increase be limited by valve masking?

These questions can only be answered by looking at how much potential increase there is in airflow, how much additional valve masking there will be, and how much of the additional masking can be eliminated. These questions, in turn, can only be answered on an engine-by-engine basis. Besides, many cylinder heads will not even allow for larger valves.

Installing oversize valves requires a great deal of machine work. First, the old valve seat must be removed. If the seat is removable, installation can be accomplished by carefully drilling or cutting through the old seat. This will relieve the tension on the seat and allow the seat to fall out of the head. If the old seat is an integral part of the head, it must be machined out.

Once the old seat is removed, the head must be readied for a larger seat. First, a counterbore is machined into the head where the old seat was. This counterbore must be 0.003 to 0.006in smaller than the outside diameter of the replacement seat. The cylinder head must then be stretched around the seat. Place the seat in the freezer overnight. The next day, place the head in the oven and bake it at 350 degrees Fahrenheit for one hour (I

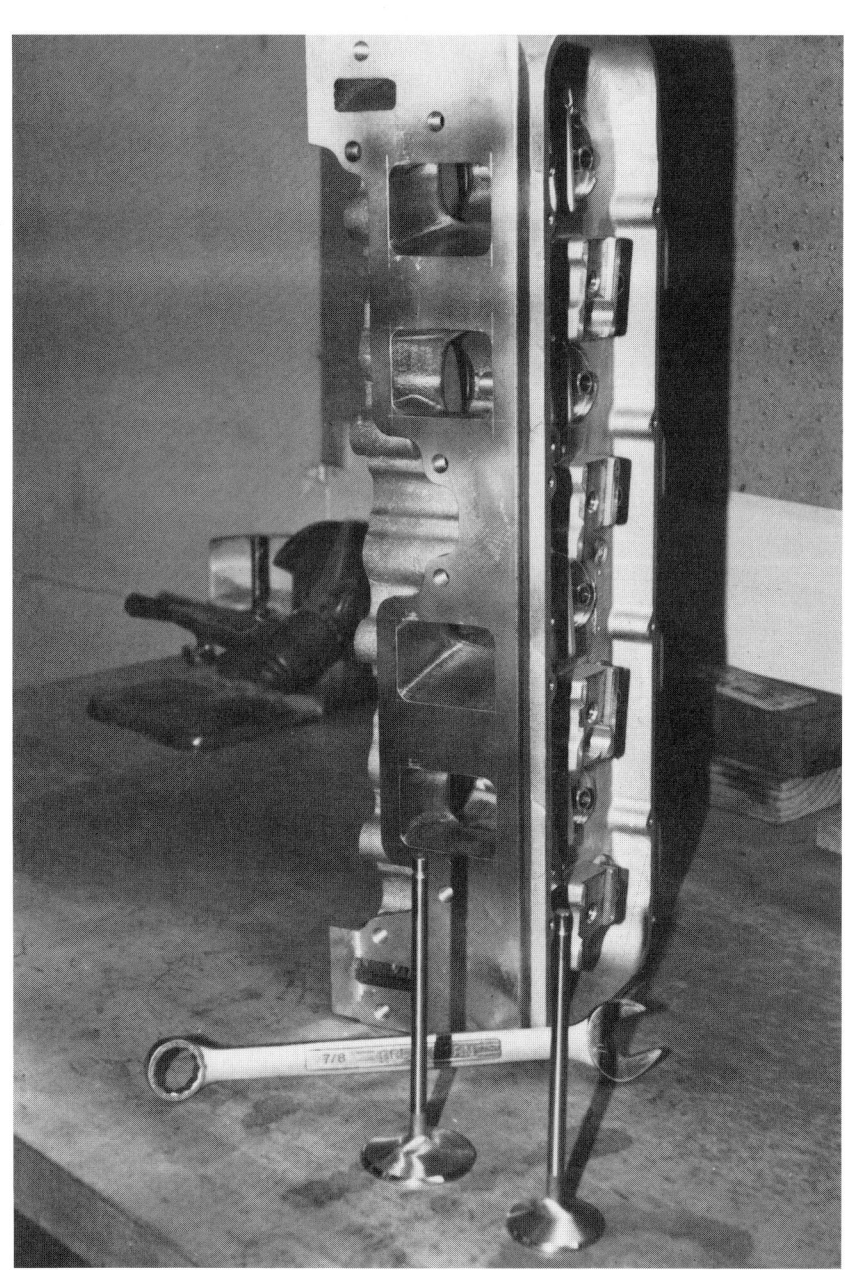

This head is itself the definition of major refitting. Oversize valves, sintered valve seats, stainless steel valves, bronze valve guides, ported polished and ground for optimized and equalized air flow and compression.

am serious!). Remove the head from the oven, remove the seat from the freezer, and drop the seat into the head.

This technique works great when only one valve seat needs to be replaced. Since the head cools quickly, you probably will have to reheat the head several times to replace all the seats using this method. When the head cools and the seat warms up, there will be an interference fit that will hold the valve firmly in position. Many machine shops will also peen the head over the valve seat. In reality this does little to hold the seat in place, but it makes many customers feel more secure.

Once the new seat is installed it should be machined with a three- or five-angle grind, as discussed earlier. If you perform the operation according to these instructions, at this point you will notice that there is a large lip

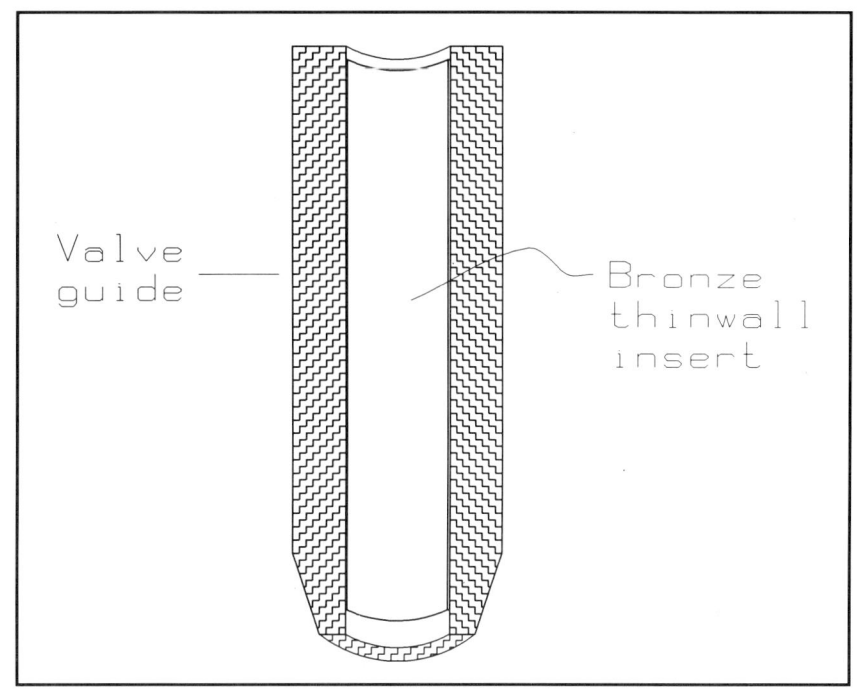

Thinwall inserts resemble a very thin bronze tube. After the existing guide is drilled and reamed oversize the insert is carefully pressed into position. Since the insert is extremely thin, this is not a job for the faint-hearted or the clumsy.

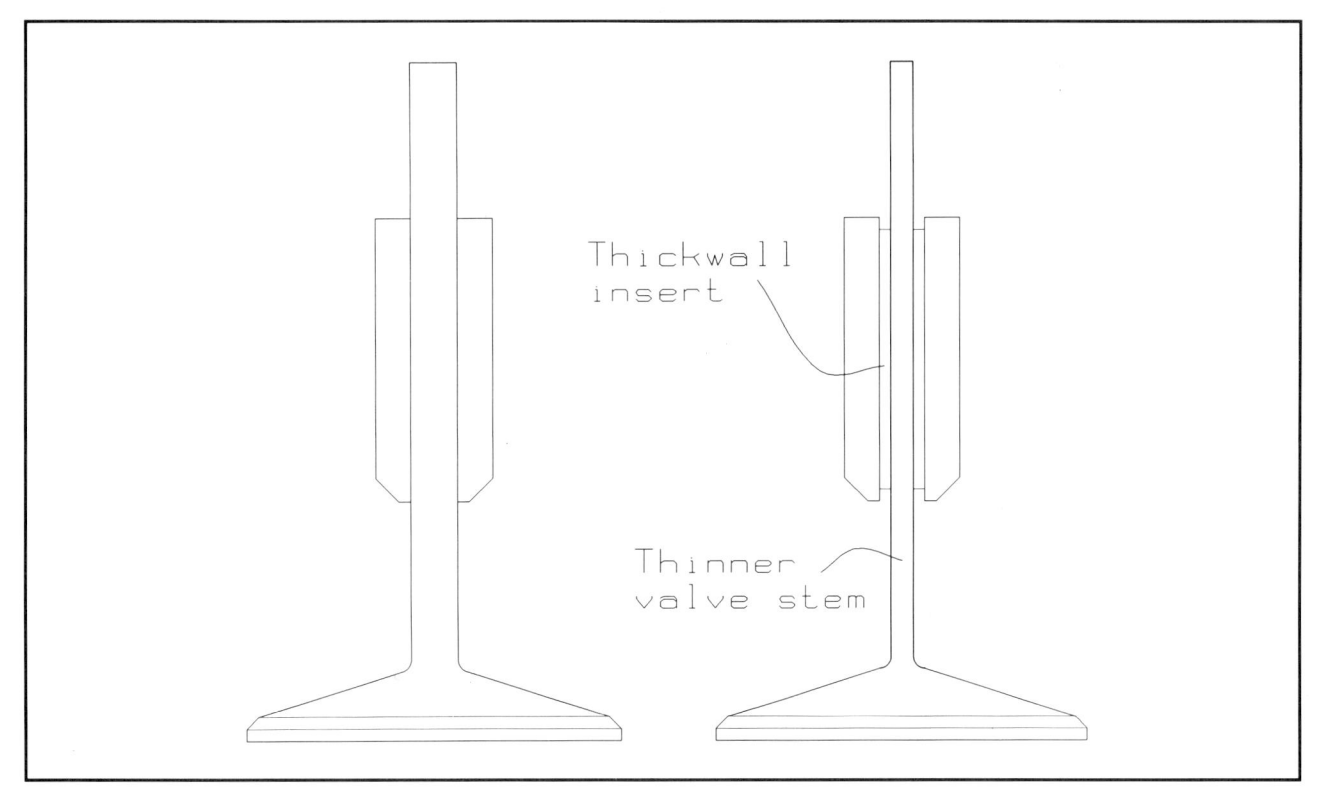

Thickwall inserts are used to reduce the diameter of the guide so that the head might be fitted with valve which have thinner stems. The thinner stems offer less interference to air flow and therefore potentially better performance.

where the valve seat heads toward the valve pocket. The valve pocket entrance must now be enlarged to taper into the valve seat and to ensure that the potential benefit of the larger valve can be realized. However, enlarging the valve pocket increases your chances of grinding though the head into a water jacket or valve pocket.

In all honesty, the benefits that you can gain from larger valves seldom justify the risk and expense. For the typical modified engine, a 350 Chevy, a 302 Ford, there are off-the-shelf cylinder heads available that provide the larger valves with intake ports, exhaust ports, and valve pockets designed to work effectively with the larger valve. When available, these heads provide a much better platform for starting head modifications than does using and so extremely modifying a stock head. Then again, there are those among us who believe that *anybody* can modify a Chevy or Ford, and that *anybody* can get superior performance out of a Dodge 340. Let's modify a Pinto 1600!

People like us are often locked up for being dangerously masochistic.

Valve Guide Inserts:
Iron

Iron valve guides are usually no more than holes drilled in a cast-iron head. Some manufacturers have used cast-iron guides in aluminum heads. Cast iron makes a fairly sturdy guide. One problem with the use of cast iron, however, is illustrated by the following story.

In the late 1970s, I was working as a technician in a Mercedes dealership. A customer came in with a 450SE whose engine was making a mild knocking sound that seemed to be coming from the top end; in fact, my first thought was valve adjustment. One of the other mechanics, Henry, older and wiser in the ways of Mercedes, said that the valve guides were loose in the heads. Doubtful, I pulled off the valve covers and inspected them carefully. Sure enough, Henry was right. The interesting part is that the customer chose not to repair the problem. Instead he went to the trunk, pulled out a paper bag, went to the showroom, and paid cash for a new Mercedes!

Bronze

There are times when a cast-iron valve guide will not do the job. High-lift camshafts, for instance, put a great deal of lateral stress on the guides. The cast-iron guide will wear quickly when a high-lift cam is installed. Aluminum bronze is an alloy commonly used in high-performance applications.

Bronzewall

These are installed in two different ways. Winina makes an insert that installs similarly to a helicoil. The guide is precision drilled oversize, threaded with a self-piloting tool, then the guide is screwed into position. After being staked into place, the guide is reamed to the proper guide to stem clearance.

A second type, such as the thinwall inserts made by K-line, resembles a very thin bronze tube. After the existing guide is drilled and reamed oversize, the insert is pressed carefully into position. Since the insert is extremely thin, this is not a job for the faint-hearted or the clumsy.

Thickwall

Thickwall inserts are similar to thinwall inserts. However, thickwall inserts are used to reduce the diameter of the guide so that the head might be fitted with thinner stemmed valves. The thinner stems offer less interference to airflow and therefore potentially offer better performance. The increase in performance is best measured in hundredths of a second at the end of a quarter mile.

Milling the Head

Generally, the cylinder head is milled only to make the mating with the cylinder block flat. Milling the head increases the compression ratio, therefore the head can be milled deliberately to increase the compression ratio. Use the formulas at lower left.

Let's say that the current compression ratio of our engine with a 3in stroke is 9.5:1 and we wish to raise the compression ratio to 11:1.

Amount to Mill =
[(10 - 8.5) (10 - 8.5)] x Stroke
Amount to Mill = 1.5 / 85 x 3
Amount to Mill = 0.052941

To raise the compression ratio to the desired level, you would need to mill a little over 0.050in off the head.

Setting Spring Pressure

Ignoring the tendency of pistons to disintegrate when their maximum rated velocity is exceeded, the single most limiting factor to engine rpm is valve float. This occurs when the cam's toe attempts to begin to lift the lifter or follower before the spring has had time to close the valve from the last opening. The greater the spring tension, the faster the valve will be closed. The greater the spring tension,

---

Compression Ratio = Displacement Ratio + 1
Displacement Ratio = Cylinder Volume / Chamber Volume
Amount to Mill = [(New Displacement Ratio - Old Displacement Ratio) / (New Displacement Ratio - Old Displacement Ratio)] x Stroke

For performance use these rockers are wholly inadequate. These rockers are mounted on studs and held in place by nut and a semi-hemispherical washer called a pivot ball. There is nothing to offer lateral stability. A high lift cam combined with high-tension valve springs can exert extreme forces on the rocker arms. I have seen the rocker arm pulled off the stud, and I have seen the pushrod perforate the rocker arm.

the faster the cam and lifters will self-destruct. The spring tension, therefore, is a compromise between the speed at which the valve will close and the amount of punishment the rest of the valvetrain can take.

There is a tool specially designed to measure spring tension. One of the more common manufacturers of the tool is Rimac. The valve spring is placed in the tool. A lever is moved to apply tension to the spring. The tension should be carefully measured at installed valve height and again at the spring height when the valve is fully open. The valve open height of the spring can be calculated by subtracting the net valve lift from the length of the valve from the seat to the bottom of the valve retainer. Compare these measurements to the specification provided by the engine or camshaft manufacturer. If either measurement is below spec, there are two alternatives.

The cheapest way to correct a below spec spring is with hardened valve spring shims. Commonly available in thicknesses of 0.015, 0.030, and 0.060in, these shims can be added in combinations to get the tension specs up to where they belong. A better way to correct the problem is to replace the springs. After replacement, shims can be used to balance the tension of the springs. (*Note:* At least one shim must be used under the spring of each valve on aluminum and alloy heads.)

This is also a good time to check the valve springs for binding. Compress the valve spring to the valve's fully open height. Check this measurement carefully. Now compress the spring another 0.060in. If the spring is in a bind, either the spring must be replaced with one that has the correct tension and does not bind, or the net valve lift will have to be altered. For engines that are going to see extended service between overhauls, which really only excludes the most ardent and well-financed drag and sprint car racers, the figure of 0.060in should be adjusted upward to 0.100in.

Not all of us are blessed with a Rimac or equivalent machine. A rather accurate and inexpensive

121

If the head you are working on is destined for a performance engine, replace the rocker arm mounting studs.

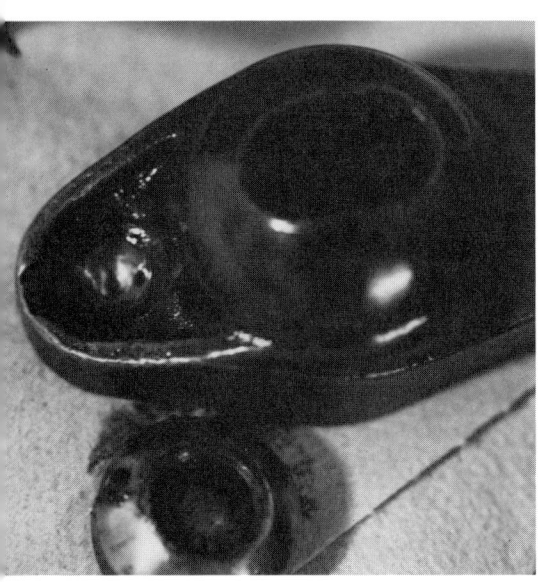

Another weak point in this type rocker is the area where it contacts the pushrod. At high speeds the pushrod can perforate the rocker.

valve spring tension tool can be constructed from a drill press and a set of bathroom scales. Place the scales on the platform of the drill press. Place a piece of plywood on the scales, then set the spring on the wood and use the drill press chuck to depress the spring. Control the pressure with the drill press lever.

Rocker Arms

There are many types of performance rocker arms available. One of the most common types of rocker arms is the stud-mounted, stamped-steel rocker. These are fairly common because they are inexpensive to manufacture, and they do an adequate job in "ma and pa" usage. For high-performance applications, however, the stud-mounted, stamped-steel rockers are wholly inadequate. These rockers are mounted on studs and held in place by a semi-hemispherical washer called a pivot ball and nut. There is nothing to offer lateral stability. Additionally, a high-lift cam combined with high-tension valve springs can exert extreme forces on the rocker arms. I have seen instances where the rocker arm pulled off the stud, and I have seen where the pushrod perforated the rocker arm.

When the engine is destined for performance use, these rockers should be replaced with cast-aluminum types. Most of these rockers come ratioed, or may offer a choice of ratios. They feature a short shaft and roller bearings; there is no need for the pivot ball. The short shaft bolts onto the stock stud, offering increased stability. The roller bearing offers increased durability. Most cast-aluminum rockers have a roller that makes contact with the end of the valve. This roller eliminates the increased wear at the contact point between the top of the valve and the rocker. The cast-aluminum design also prevents pushrod perforation.

A second type of stock rocker arm mounts the rocker on a shaft. These rockers can either be made of stamped steel or cast. Inherently more stable than the pivot-ball rocker arms, the shaft-mounted rocker arms *can* be retained when performance modifications are done. The cast rocker arm runs little risk of pushrod perforation when high-lift cams and high-tension valve springs are used. The only reason to replace these rocker arms is to decrease wear between the rocker and the shaft. When a high-lift cam or high-tension valve springs are used, replace the stock rocker arm with roller rockers.

Rocker Arm Guide Plates

Rocker arm guide plates are available for use with stud and pivot-ball rockers to keep them properly aligned. However, they are not nearly as effective at maintaining a properly aligned valvetrain as the shaft-mounted rockers.

The guide plate mounts on the stud and has a groove in which the pushrod travels. The pushrod rubs against the guide plates as the engine runs. As a result, stock soft-metal pushrods will self-destruct in a short time. Therefore, hardened pushrods should be used.

Retainer-to-Guide Clearance

One inexpensive, but easy and effective trick to increase valve lift is to use ratioed rocker arms. As the valve lift increases, whether through the use of a high-lift camshaft or through the use of ratio rockers, the valve spring retainer gets closer and closer to the top of the valve guide. Before the engine is fully assembled, insert the valves into the valve guides, holding them in place with a blob of modeling clay. Snug the cylinder in place with a head gasket installed. Rotate the crankshaft until the pushrod is at its highest point, then install and adjust the rocker. Install the retainer and valve keeper on the valve. Be careful not to drop the valve into the cylinder. The distance between the bottom of the valve keeper and the top of the valve guide should be a minimum of 0.060in. Repeat this procedure for each of the valves.

Pushrods

A number of variables can affect rocker arm geometry, includ-

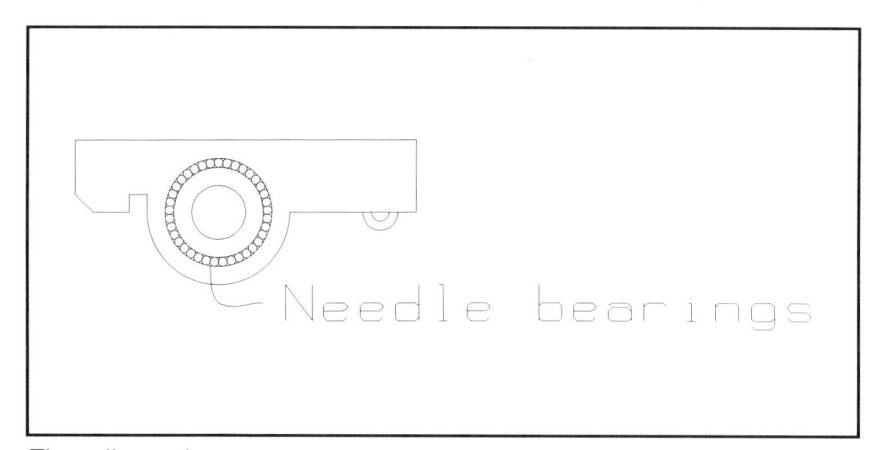

Roller rockers are made by several manufacturers and eliminate most wear and friction problems typical to rocker arms. Note: Be sure to be seated when they quote you the price compared to the price of the stock rockers. Yet they are perhaps the best investment that can be bolted on to your cylinder heads.

The roller rocker arms mount on a shaft and use needle bearings to reduce friction and wear.

For most stock applications the only thing that keeps the pushrod properly aligned is the rocker. At high speeds, especially if the speed of the engine exceeds the ability of the valve spring to close the valve (valve float) the pushrod can become misaligned.

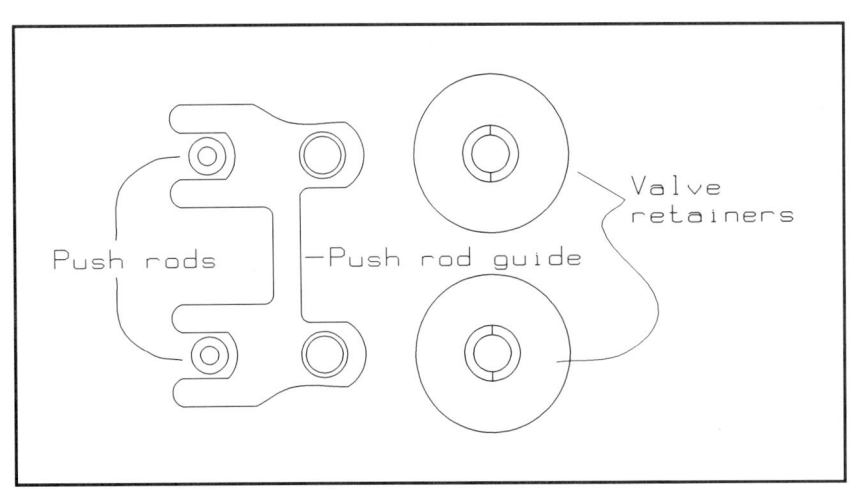

Pushrod guide plates can be purchased to keep the pushrods properly aligned when the engine is running at high speeds.

ing decking the block, using camshafts with non-stock base circles, and milling the cylinder heads. The problems created by these operations can be corrected by installing different length pushrods. Available in a variety of lengths, the correct length can only be found through trial and error.

One of the drawbacks to using ratioed rockers is the possibility of the pushrod rubbing against the side of the hole where it passes through the head. As you begin to adjust the valves, be sure to check for rubbing or binding.

Valve retainer clearance

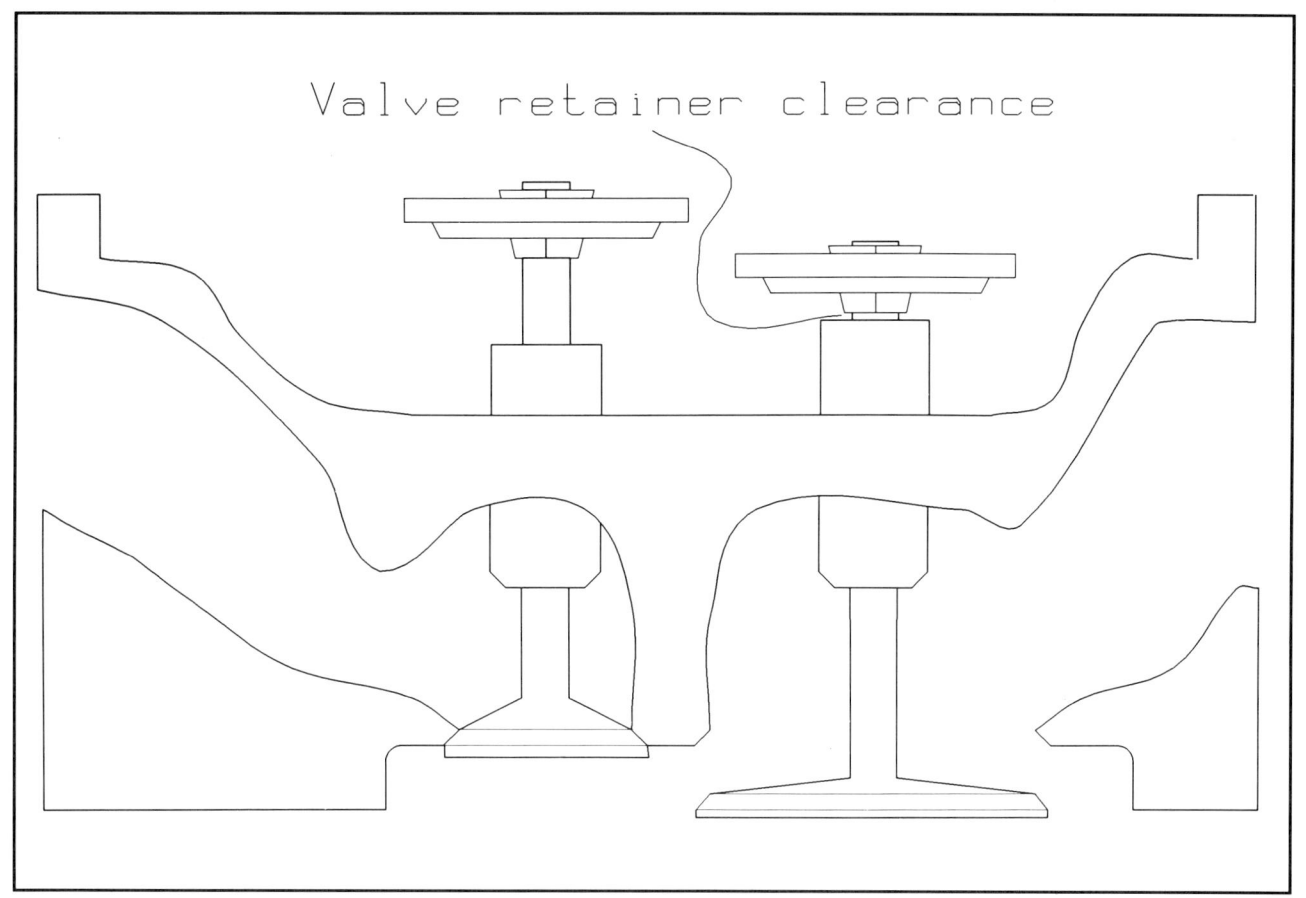

Before the engine is fully assembled, insert the valves into the valve guides holding them in place with a blob of modeling clay. Snug the cylinder in place with a head gasket installed. Rotate the crankshaft until the pushrod is at its highest point, install, and adjust the rocker. Install the retainer and valve keeper on the valve. Be careful not to drop the valve into the cylinder. The distance between the bottom of the valve keeper and the top of the valve guide should be a minimum of 0.060in.

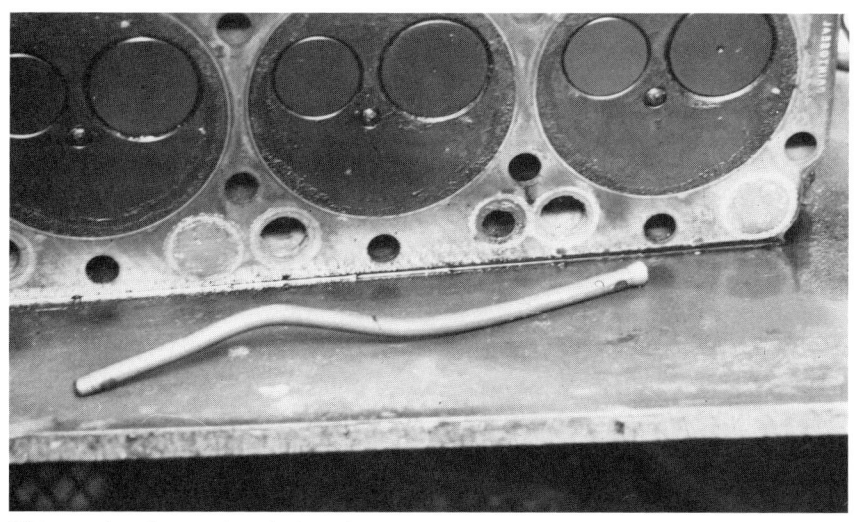

This pushrod was the victim of poor maintenance. The valve was improperly adjusted, the pushrod slipped out of the rocker, the pushrod and the head were destroyed. Pushrod guides might have prevented this, or proper maintenance would have prevented this.

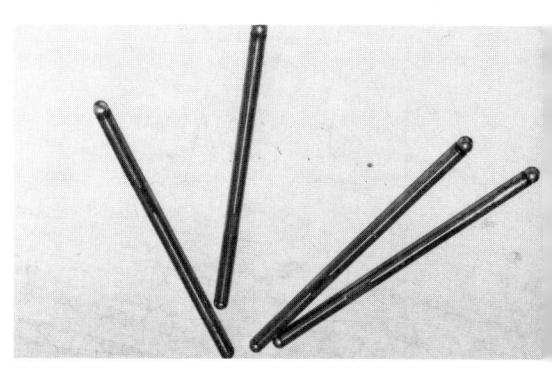

Pushrods are available in a variety of lengths, The correct length to cure the effect of these machining operations can only be found through trial and error.

Engine Reassembly, Camshaft and Valve Adjustment

13

Begin engine reassembly by sliding the valves into place. Now install the valve seals. Some valve seals snap onto the valve guide and are held in place through friction. These valve seals are popular on domestic engines. Another type is the umbrella seal. They just slip onto the valve stem, forming an umbrella over the top of the valve guide. Umbrella seals are also popular on domestic engines. Do *not* think that new valve seals will cure oil burning resulting from worn valve guides. Valve guide repair is the only remedy to that problem.

Most top end gasket or valve seal kits come with a small Mylar sheath that fits snugly over the end of the valve stem to protect the seal as it slides over the stem. Failure to use the sheath can result in a cut seal and an oil-burning problem.

With the valves and the valve seals in place install the valve springs, retainers, and keepers. The valve spring compressor you used to take the head apart needs to be used again to reassemble it.

If the engine is an overhead cam design, install the camshaft now. Some applications use a bucket or shim cam follower. First, check for free rotation of the camshaft. The followers should be installed after checking and removing the cam. The installed followers would bind the camshaft, preventing it from rotating freely. If the cam does not rotate freely, it will be necessary to have the cam journals line bored as described in the machining chapter.

With the free camshaft rotation confirmed, remove the camshaft if necessary to install the cam followers. For most applications that use a rocker type follower, the rockers will slip into place when the valve is partially depressed. Many professionals line up the rocker and drive it into position with a hammer. While this may not do any obvious damage to the rocker, cam lobe, or the rocker support, it can damage the hardening and cause premature component failure.

Cylinder Head

This is the part of engine reassembly that requires the best technique and the most care. Be sure both the head and the block mating surfaces are clean. Place the head gasket on the block deck. Look carefully to ensure that the proper side is up (head gaskets are not always marked for proper installation). The ports in the water jacket and any ports in the oil gallery in between the cylinder head and block should line up through the head gasket. If the head is properly machined, flat sealers are not needed. If the head is not properly machined or flat, it should be repaired or replaced.

If the cylinder head was resurfaced on the bottom, the combustion chamber volume has been reduced. This raises the compression ratio. If the engine has a 4in bore, removing 0.030in from the head will decrease the combustion chamber by 0.4cc. This would raise the compression ratio by only 0.5 percent or so. However, if you are building a relatively high compression engine that you plan to run on pump gas, you may want to consider using a thicker than stock head gasket, or if available, a shim.

Place the cylinder head gently on the head gasket. Be careful not

Before installing the head make sure the block has been properly prepared. It makes no sense to spend hundreds or even thousands of dollars performance-preparing the heads and set them on top of a poorly prepared block.

Take a little time and prepare the water ports and oil galleries of both the head and block.

Which way is up? On many engines the head gasket is almost symmetrical. If not installed properly, however, serious damage can occur to the head and engine.

Almost without fail the head gaskets are marked for front and top. This gasket should be mounted with the word "front" visible and toward the front of the engine.

to allow the edges of the head to dent or otherwise damage the gasket. Many head gasket failures are a result of damage that occurred during head installation.

Now comes the most critical part of the head installation process: torquing. In general, cylinder heads are torqued from the center to the ends in progressive steps. This prevents warping the head. Some heads are torqued to an angle rather than the more common foot/pound specification. The most notable of these is the Volvo V-6 engine. If possible, consult the factory service manual for the proper torque spec. If no service manual is available, it is usually acceptable to begin with the center-most bolt, torque it to 25 percent of the spec, go to the bolt the next clos-

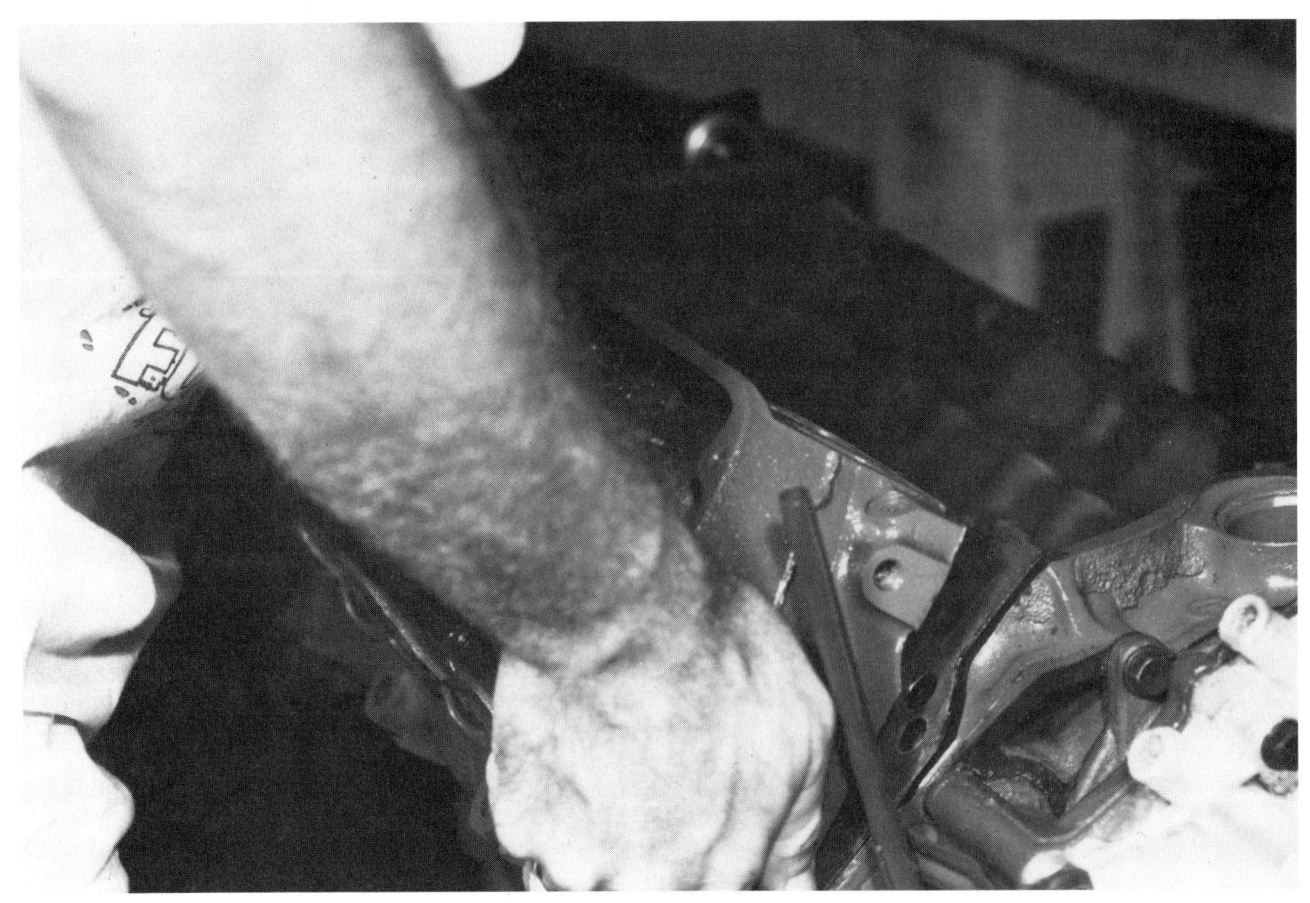

Set the cylinder head on the engine block carefully. What may seem to be an inconsequential nick at this point may cause an engine failure 20 feet past the starting line.

128

```
10    3    1    5    7

 8    6    2    4    9
```

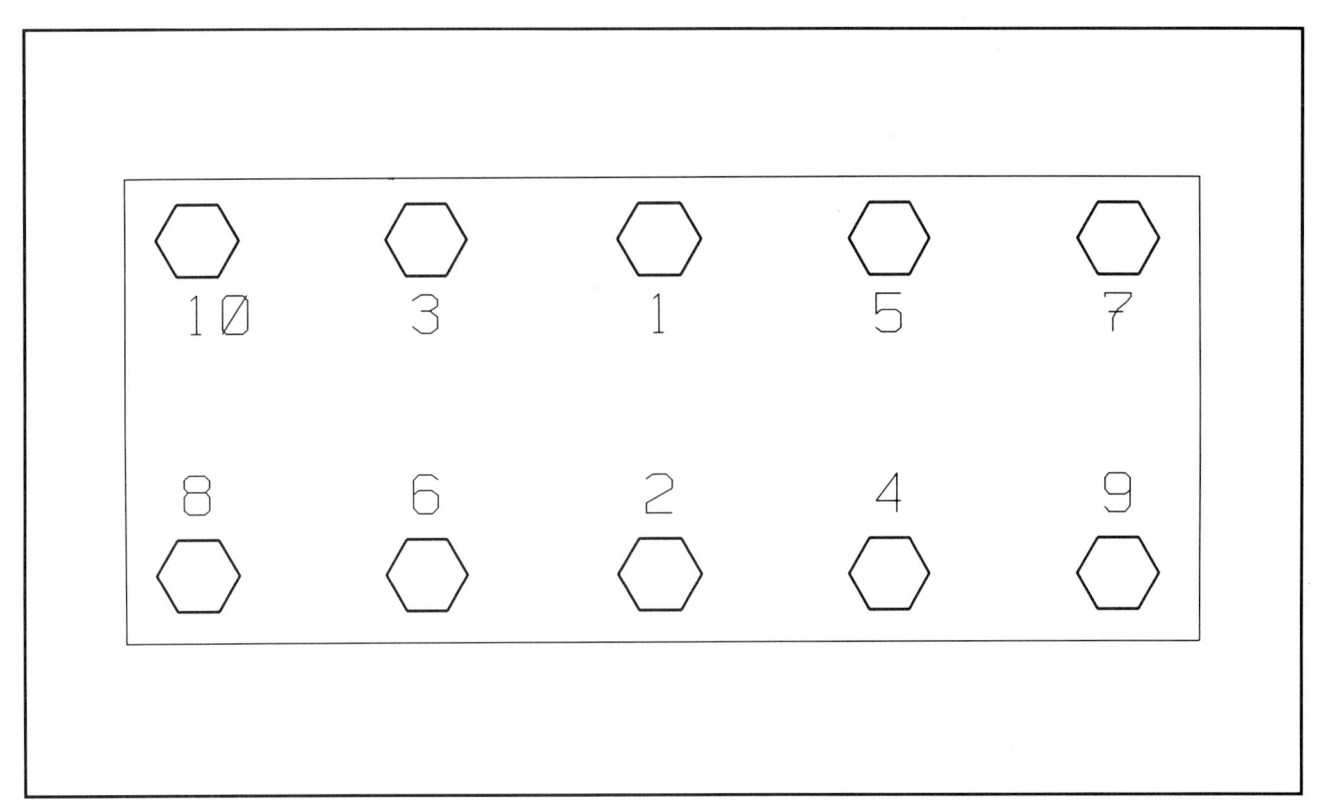

Generally cylinder heads are torqued from the center to the ends in steps. This prevents warping the head. Some heads are torqued to an angle rather than the more common ft/lbs (foot/pound) specification. The most notable of these is the Volvo V-6 engine. If possible, consult the factory service manual for the proper torque spec. If no service manual is available use the pattern illustrated above.

est to the center, and begin tightening the bolts in an hourglass pattern, criss-crossing the head until all are torqued to 25 percent of the spec. Repeat this process at 50 percent of the torque spec, then 75 percent, and finally 100 percent. On applications with aluminum heads, it will be a good idea to retorque the head after the engine has been run for a few hours and allowed to cool.

Overhead Camshaft Timing

When I owned a repair shop in Bellevue, Washington, I was asked to do a tune-up and valve adjustment for the son of a friend. If you ever own a repair shop, *never* do a tune-up and valve adjustment for the son of a friend. The car was a Honda 600. While adjusting the valves, I had the transmission in gear so I could roll the car back and forth to bring each pair of valves up to adjustment. At the time I did not know that the timing chain could skip a tooth if the engine was rotated backwards. I rotated the engine backwards, and the timing chain jumped. Since the valve timing was now off, the valves, which had very thin stems, were bent.

If the engine is an overhead cam design, now is the time to adjust the cam timing and install the timing chain or belt. If a timing belt is used, be aware that many timing belts are designed with a front side and a back side. These belts should be installed only one way. Consult the factory service manual for details.

Overhead Valve Lifters and Pushrods

Install the lifters. Do not use new lifters on an old camshaft, and do not install a new camshaft without new lifters. Put a little assembly lube on the sides and especially on the bottom. The lifters should fall into place with only the slightest pressure.

Set the pushrods through the heads and into the lifters. Then install the rocker arms.

Valves

Almost all of today's engines have hydraulic lifters or hydraulic valve compensators. They will probably require readjustment after a few minutes of running. No matter if the lifters are hydraulic or not, they should be adjusted at this point.

Mechanical Lifters

Mechanical lifters are generally adjusted to a hot specification. This information can be found on the EPA sticker located under the hood. If no cold specification is given, adjust the valves

129

Always check with the factory specifications when setting cam timing. In general, most applications will have the marks on the cam and crank gear directly opposite one another. If you followed the procedure mentioned earlier in the book for degreeing or "phasing" the cam, be sure and use the Woodruff key determined to be appropriate by that method.

to 0.002in tighter than the hot specification.

Rotate the crankshaft and observe the lifters on the companion cylinder. When one of the lifters on the companion cylinder stops moving down and the other lifter starts moving up, the valves on your cylinder are ready to adjust.

The companion cylinder is the one opposite in the firing order. Check the factory service manual for the firing order to the engine you are working on. Some typical firing orders are shown in the chart at right.

To find the companion cylinder easily, divide the cylinders at midpoint: the first on the first half of the cylinders is the companion to the first on the second half of the cylinders; the second on the first half of the cylinders is the companion to the second on the second half of the cylinders, and so on. In the V-8 engines with the firing order 15486372, 1 and 6 are companion cylinders, so are 5 and 3, 4 and 7, and 8 and 2.

With the lifters of the companion cylinder rocking, insert the correct thickness feeler gauge between the rocker arm and the top of the valve stem. Tighten the rocker arm against the feeler gauge until it slides snugly in and out with slight resistance. Repeat this process for each of the cylinders.

Hydraulic Lifters

The preliminary process of adjusting the valves is the same for overhead valve engines with hydraulic lifters as for those with mechanical lifters. However, final adjustment of the lifter must be made with the engine running.

Inline Engines	Firing Order
4-cylinder	1342 or 1243
5-cylinder	12453
6-cylinder	153624 or
	124653 or
	142653 or
	145632
8-cylinder	16258374 or
	13684275 or
	14738526 or
	13258674
V Engines	**Firing Order**
4-cylinder	1324
6-cylinder	125643 or
	145623
8-cylinder	16354728 or
	15486372 or
	18364527
Pancake Engines	**Firing Order**
4-cylinder	1432

New lifters, or even old hydraulic lifters that have bled down during the rebuilding or cleaning process will compress easily as the rocker is adjusted. This would allow the valve to open too far after the engine is started and the lifters pump up. There is a potential for damage when the valves are in overlap on the transition between the exhaust and intake strokes. For a moment, the intake and exhaust valves are both open as the piston is approaching the top of its stroke. Should the lifters be adjusted too tight after they pump up, the valves may meet the piston. It is devastating to an engine to have two metal objects attempting to occupy the same space at the same time. To ensure the lifters will not cause damage to the valve once the engine is started, be careful not to depress the lifter as the valve is being adjusted.

Overhead Cam

In many cases, the machinist will have made preliminary adjustment of the valves when doing the grind job. Be sure to ask about this when you pick up the cylinder head. Different ohc configurations will require different adjustment techniques. The three most common configurations for ohc valve adjustment are: rocker arm, shim on top of the follower bucket, and shim under the follower bucket.

The adjustment procedure for the rocker arm design is exactly the same as for the overhead valve engines. Rotate the crankshaft until the valves of the companion cylinder are in overlap. Note that the heels of the cam lobes on the cylinder you are adjusting are both in contact with the rocker. Adjust the valves with the adjusting screws at the ends of the rockers.

For applications where the shim is located on top of the follower bucket, a special tool is required to depress the follower and valve spring to change the shim.

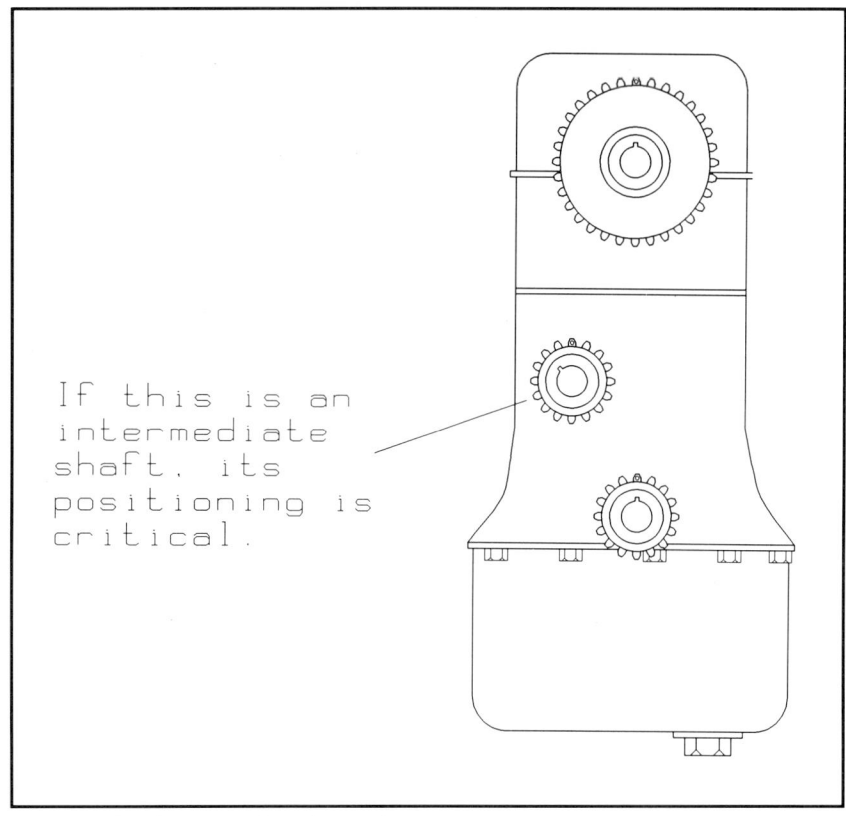

If the engine is an overhead cam design, now is the time to adjust the cam timing, and install the timing chain or belt. If a timing belt is used, be aware that many timing belts are designed with a front side and a back side. These belts should only be installed one way. Consult the factory service manual for details.

This tool slips between the camshaft and the follower, depressing the follower as the tool is levered downward. Measure the clearance for each valve. If the valve clearance is incorrect, remove the old shim using the special tool and replace it with a shim that is the correct thickness. These shims are easy to come by from the dealer, and are even available after-market. The thickness of the shims is usually in increments of tenths of a millimeter. In many cases, the old shim may be difficult to remove from the follower. Although a special pair of pliers resembling a snap ring plier can be purchased, a pick generally works just as well.

Another type of adjustment shim design places the shims under the bucket. Replacing this type of shim requires removal of the camshaft. The easiest way to adjust these valves is to check the adjustment of all the valves, determine how much of a change needs to be made on each to remove the camshaft only once, and replace the shims. As you reinstall the camshaft after replacing the shims, make sure the valve timing is correct.

With the valves adjusted, install the oil pan, the intake manifold, the valve covers, the exhaust manifolds, and the flywheel or automatic transmission flexplate.

Oil Pump

Remove the cover plate on the oil pump. Pack light grease between the gears of the pump. This will prime the pump and get oil to the bearing surfaces as quickly as possible on initial startup.

Install the lifters gently, do not "slam" them into place as this may cause the hardening on the bottom of the lifter to be damaged or dented.

Oil Pan and Valve Covers

The only thing that is not obvious about installing the oil pan and valve covers is the occurrence of gasket dimples. Gasket dimples form as the oil pan or valve cover is bolted into place. The stamped-steel units tend to dimple as they are torqued. Check for this and repair by peening from the backside before installing.

Intake Manifold

Set the intake manifold gaskets in place. If this is a V engine, the gaskets may not want to stay in place as the manifold is set into position. Applying a light film of grease to the gasket surfaces will solve this problem. Torque the manifold in a decreasing hourglass-shaped pattern starting at the outer ends. Torque in 25, 50, 75, and 100 percent stages, just like for the cylinder heads.

Distributor

Installing the distributor may be the trickiest part of the engine reassembly process. Set the crankshaft to top dead center number one compression. Drop the distributor into place so that when the distributor is in the seated position, the rotor points to the number one position on the distributor cap. While this does not ensure that the ignition timing is correct for driving the car, the timing will be set close enough to start the engine. The gear at the bottom of the distributor has a helical cut. When the distributor slides into position, the distributor shaft will rotate the equivalent of about two teeth. Therefore, the teeth of the distributor gear should be aligned about two teeth out of phase as you begin to drop it into position.

Exhaust Manifold

It is a good idea to make sure the exhaust manifold is flat and true before installing it. Many applications depend on a machine fit between the exhaust manifold and the head, rather than using a gasket. Torque the manifold in place from the ends to the center in stages.

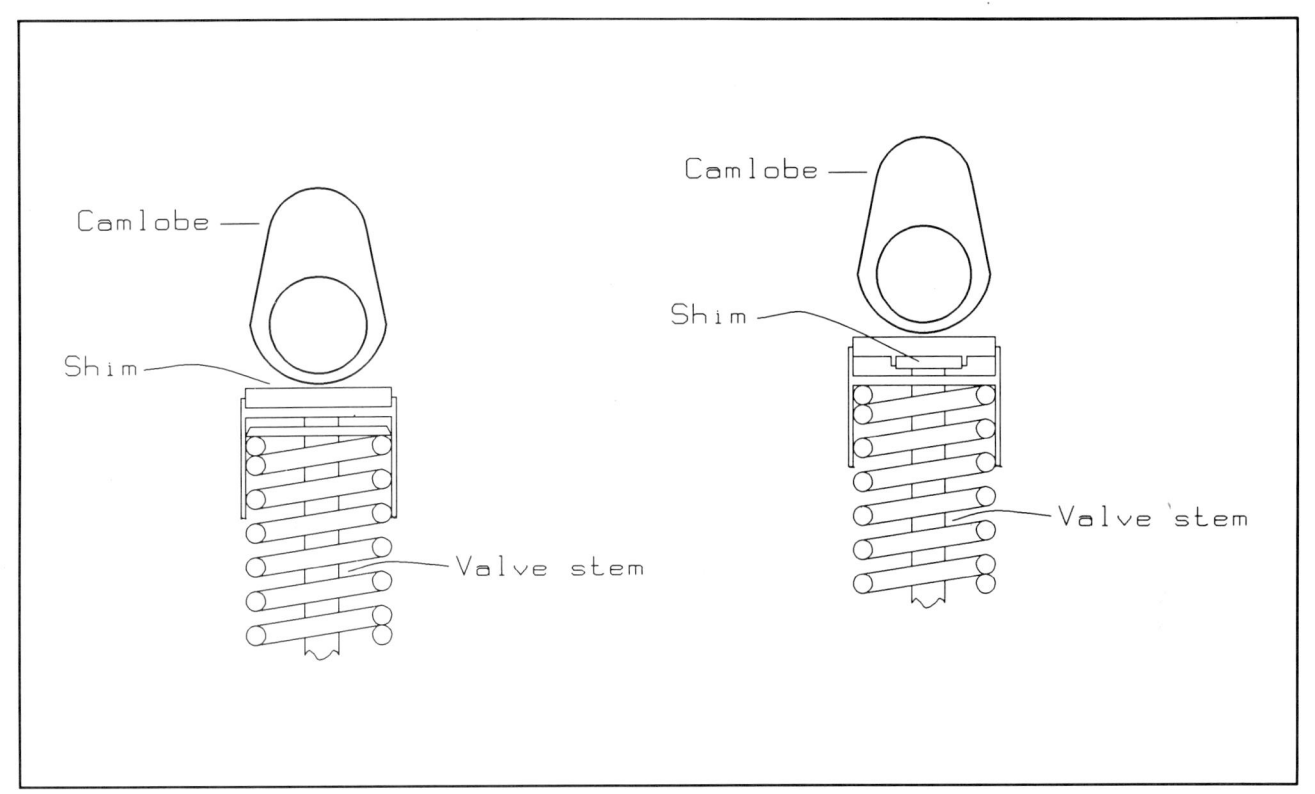

Overhead cam engines that do not have rocker arms adjust the valve using shims. These shims might be located above or below the cam follower. Volkswagen and Volvo products generally place the shim on the top of the follower while SAAB places it below. On the SAAB application the camshaft must be removed each time a shim is replaced. For applications such as these all the valve adjustments should be measured at once so all the shims can be replaced the first time the camshaft is removed.

This may be the trickiest part of the reassembly process. Set the crankshaft to top-dead-center number one compression. Drop the distributor into place so that when the distributor is in the seated position the rotor points to the number one position on the distributor cap.

133

Other Performance Modifications

14

Nearly twenty years have passed since the golden age of the Detroit performance car. For nearly two decades, concerns about emissions, fuel economy, and safety have condemned the enthusiast to either driving late-model slug-bait or rebuilding wrecking yard refugees. However, beginning in the latter part of the 1980s and continuing in the 1990s, the late-model sleeper is again a realistic possibility. The manufacturers themselves have shown a renewed interest in performance, starting with the 1984 Buick Grand National. In 1987, Oldsmobile introduced the Quad 4 engine. This was a four-valve-per-cylinder 2.3 liter four-cylinder engine that produced 154hp stock—more than 1hp per cubic inch. Companies that were performance names in the sixties, along with a whole slew of new manufacturers, today are hitting the market with performance parts.

In writing this section on performance modifications, it comes to mind that there is one major difference between modifying engines back in the 1960s and modifying engines today—emission laws! Different jurisdictions will have different mandates concerning emission control systems and what modifications are legal. Generally speaking, throughout the US it is illegal to remove or modify any emission control device on a car that will be licensed

I have very mixed emotions about the air cleaner. For stock applications the factory air cleaner is capable of flowing more air than the engine will probably ever require. Yet when the heads have been modified and the camshaft changed, the stock air cleaner should be replace with one of the many "performance" models which are available. Be sure to confirm with the local authorities that the air cleaner you buy is emission legal.

for street use. There is, however, a small amount of latitude permitted in some states for the replacement of some components like intake manifolds, throttle bodies, and camshaft with specific tested and approved parts.

This chapter, therefore, will be divided into two major categories: street modifications and racing modifications. Please keep in mind that regulations change frequently when it comes to emission controls, so be sure to ask your part supplier if a given modification is street legal in your area. Do *not* accept this book as the final word on what may or may not be legal in your specific jurisdiction.

Intake System Modifications

For street use, many of the modifications to the intake system that were traditional in the 1960s were already done by the manufacturer when the engine was designed. A simple fact is that any meaningful intake modification will be found in the air cleaner and ducting to the throttle body, or in the complete rethinking and reworking of the intake side of the engine including throttle body, intake manifold, and valvetrain.

Let's begin by looking at what you already have on your car, and discuss what can be done to improve it. The example we are going to be using is the 5.7 liter Tuned Port Corvette. As we cover each of the above-mentioned topics we will look at the availability of over-the-counter performance upgrading parts for not only the 5.7, but also for the 5.0 and 2.5 liter versions.

Air Cleaner Assembly

The air cleaner is just one component in the total air induction system. For a street application it would be a real shame to spend hundreds or thousands of dollars and dozens of hours upgrading, refitting, porting, and polishing the heads only to have the work be ineffective because of restrictions in the air cleaner or air filter.

During normal operation, the air cleaner on most engines is capable of handling more than the specified airflow. However, if the car is to be used under circumstances where maximum performance is required, such as racing or trailer towing, then the air cleaner assembly can become a restriction to airflow and therefore performance.

In the old days, car enthusiasts would flip the air cleaner lid upside down. While the effects of doing this were largely psychological, it did bypass the snorkel tube of the air cleaner and greatly increased the potential airflow. The design of fuel injection air cleaners does not make inverting the lid practical or desirable.

According to studies, the air cleaner assembly belonging to a late-model 350 Corvette, with the stock lid and stock air filter, is capable of flowing 648cfm (cubic feet per minute). This seems to be a comfortably above the 426 estimated cfm for the 350 Tuned Port engine and indeed a modification on a stock engine would not be necessary. In fact, there is enough airflow capability in the stock air cleaner to handle 6400rpm at 100 percent volumetric efficiency. What this means to the average motorist is that the air cleaner will become a factor only when there is a desire to suddenly change the pressure in the intake manifold.

When the throttle plates are opened, the first thing that occurs is a sudden increase in the mass velocity of the air entering the intake manifold. Cruising at 2000rpm, our 350 is only swallowing about 141cfm. When the throttle is matted, the demand suddenly jumps to a potential of 434cfm and as the rpm climbs, so will the demand. The more open and free flowing the intake system is ahead of the throttle plates, the faster the air mass will build and therefore the faster the torque will build.

Several companies make air cleaner housings for a wide variety of applications that boast of increased cfm capability and therefore increased ability to make a sudden change in mass airflow rate. A crude, but effective method of accomplishing the same thing is to trim the air filter lid so that all of the filter is exposed and the airflow through the filter is totally unrestricted by the cover.

The air filter itself can severely impede airflow. High-performance air filters are available for a wide range of applications. These filters cost about $50, but are one of the easiest—and legal—ways of improving power.

How do these performance air filters provide a noticeable improvement in horsepower over the standard air filters? To answer that question, we need to begin with the premise that a stock air filter is designed to be easy and inexpensive to produce. The consumer who is not performance oriented is interested in reducing the cost of maintenance, and these stock air filters are designed to meet that need. A performance air filter, on the other hand, sacrifices low price for increased airflow.

The high-performance air filter consists of multiple layers of cotton gauze that has been treated with a light oil. The oil attracts dirt and dust particles, and the loose weaving of the gauze allows for more effective filtering. Typically, the filtering element will be layered across a metal grid which holds the gauze in an accordion shape. As the gauze weaves its way back and forth, the frontal area where air passes into the engine is greatly increased. High-performance filters can result in a 25 percent increase in cfm through the filter. Keep in mind, though, that a filter capable of flowing 500cfm will

This is the insides from a Bosch style Mass Air Flow sensor. Note that the air mass measuring element and the cooling fins can severely limit air flow.

not increase performance significantly if it is mounted in an air cleaner assembly that will only flow 200cfm.

Some of the sales literature concerning performance air filters states that they are not legal for use on California pollution-controlled cars. The 1990 Bureau of Auto Repair (BAR) *Smog Check* manual states in Appendix K that the air filter is a Category 1 component. Category 1 replacement parts are not considered to be of concern as long as none of the emission control devices themselves are tampered with during installation. Check with the shop that does your smog check about their interpretation of this rule, or contact your local BAR Field Office for a ruling.

The $50 air filters are designed to last a lifetime. They do need to be serviced on a regular basis, however. A special solvent or cleaner is sold which removes road oil and dirt. Your new air filter will come with recommendations on the correct cleaners to use. After cleaning, the oil barrier must be replaced. Using the wrong oil, such as penetrating oil, and so on, even though the oil *looks* right can reduce the efficiency of the air filter to the point where you would be better off with a stock filter. Again, follow the recommendations of the filter manufacturer. And plan on cleaning the air filter every time you change the oil. This may be a little more frequent than is required, but it can't hurt.

Mass Airflow Sensor

Several late-model applications use a mass airflow sensor (MAF) to measure the amount of air entering the engine so that the computer can inject the correct amount of fuel. The MAF is one of the worst offenders in the restriction of airflow. The one used on the Tuned Port Corvette was designed and built by Bosch. Studies indicate that it is capable of flowing only 529cfm stock. There are screens located at each end of the sensor. These screens were put there to protect the delicate heated wire that is suspended across the main channel of airflow. Flowbench testing indicates that an increase of 182cfm, up to a very respectable 711cfm, can be obtained by simply removing these screens.

The MAF also has cooling fins for the electronic module on the side of the sensor, which sit across the main channel of airflow. These fins represent a restriction to airflow of about

Newer design MAF sensors place the air measuring element out of the main stream of air flow.

One of the cheapest and most dramatic performance modifications I ever made to an engine was to replace the throttle body assembly. The difference was so dramatic, in fact, that the owner was back in a few days with his transmission bolts ripped out. Replacement, high-performance, throttle bodies are now available for many applications.

39cfm. While this is not nearly the restriction created by the screens, the cooling fins represent a potential loss of about 10 percent of the total horsepower potential of the engine. Not only do we have these fins to contend with, but we also have the hot wire sensing element which sits across the main channel of airflow. By simply doing its job, the element seriously reduces potential airflow. A rough estimate of the restriction created by the sensing element is about 20 percent. We cannot do anything about that, but we can do something about the cooling fins.

Legend has it that electronic failures of the MAF occurred during Death Valley testing. The cooling fins were then added to increase the ability of the MAF electronics to give up heat in high-temperature environments. There is probably a certain amount of truth to this, since the Bosch fuel-injected applications such as Volvo and Porsche sport no such fins.

If you are not going to be operating the car in a hot-weather environment, then grab your hacksaw, your two-hundred-mile-an-hour tape, and get busy. Cover each end of the sensing element venturi (the circle within the circle) with the tape. Ensure that the interior of the sensor is free of grease, oil, moisture, or anything else that might attract and hold the metal sawdust that is going to result from removal of the cooling fins.

If you are like me, you may feel something ranging from quiet trepidation to out-and-out fear about taking a hacksaw to a $400 electronic component. Aftermarket sources sell a stock unit to replace one that is already modified for less than dealer list. If you prefer, and can do without your car for a few days, your MAF can be professionally modified for less than $150.

Most General Motors PFI (port fuel-injected) applications use the Delco MAF. However, the design of this unit does not lend itself to modification.

Oddly, Oldsmobile on their 3800 series engines uses a Japanese-designed MAF. In this Hitachi design the hot wire sensing element is located in a bypass tube outside the main channel of airflow. This technology not only eliminates the airflow restriction created by the cooling fins, but also the restriction of the sensing element itself, leaving, in effect, a big hollow tube for the air to pass through. Unfortunately for the performance enthusiast, this MAF is not interchangeable with the Corvette MAF (that is, unless you possess an exceptional amount of free time and a degree in electronics).

Throttle Body Assembly

Another source of restriction is the throttle assembly itself. On carbureted engines this means the carburetor itself. The throttle body assembly or carburetor holds possibly the most important potential for "bolt-on" improvement short of the intake manifold.

Several companies manufacture high-performance throttle assemblies, but let's begin our look at throttle body modifications by looking at a more inexpensive and clever approach. TPIS (Tuned Port Injection Specialists) and Hypertech sell an airflow directing device that helps to guide the incoming air through the throttle bore and into the intake manifold. Called an airfoil, this unit simply bolts onto the throttle assembly and adds an additional 8-10hp. Documentation on the 350 'Vette engine shows an increase of up to 13lb-ft of torque at 4500rpm. Fine-tuning the flow of air through the throttle valve assembly with a device such as the airfoil increases cfm in a flowbench test by 17cfm.

Big bore kits are available from Air Sensors, Holley, and others. These kits consist of a complete throttle body assembly which replaces the stock assembly. The Air Sensors version features a single large throttle plate, replacing the small throttle plate pair in the stock unit.

Holley markets a wide range of TBI replacement units. These units fall into two general categories: stock emission legal/street performance and street competition/non-emission.

The emission-legal units are available in a 300cfm throttle body assembly for the 2.0 and 2.5 applications, and a 400cfm two-barrel version for the 2.8 V-6 TBI truck engines. These throttle bodies are direct replacement bolt-on assemblies that come complete with the TPS and IAC. Once installed, all that has to be done is to connect the vehicle wiring harness to the TPS, IAC, and injector, adjust minimum air and TPS closed-throttle voltage, and then drive the car away. The units are for use on model years 1982 through 1986. Applications later than 1986 use a different throttle body assembly which is not interchangeable with these units.

The non-emission TBI units come with their own computer and wiring harnesses. For TBI applications, replacement of the intake manifold can mean a very noticeable improvement in performance.

When the installation of either the airfoil, the big bore kit, or replacement TBI is complete, it may be necessary to adjust the minimum idle speed. Unless specified in the installation instructions or in data supplied with other components that have been added to the engine, you should adjust the minimum idle speed to the specifications found in the minimum air adjustment charts.

Intake Manifold

Many late-model intake manifolds are designed for peak performance; however, production tolerances are not such that perfection is always achieved. These manifolds consist of a large, central plenum box connected to the intake manifold base by individual tubes called runners. Some improvement in cfm can be achieved by matching the sizes and alignment of the plenum, runners, and manifold base.

Matching plenum alignment can be achieved by applying a light film of Prussian Blue to each end of the runner, installing and torquing down the runner. Removing the runner and inspecting the Prussian Blue contact point will show where the runners are misaligned. A more careful inspection will reveal where the runners are undersize or oversize compared to either the intake or central plenum. A die grinder and an ample supply of patience will enlarge the undersize orifices to match the larger. Do not get carried away, though, because the size and shape of the plenum, runners, and manifold are all basically correct for the specific type engine.

After painstaking hours of work, you have gained only about 2 or 3hp. You might say that it is hard to improve on what is basically a good design.

There are high-performance manifolds available for applications ranging from the Honda CVCC to the basic Chevy. These manifolds offer larger porting, improved runner matching, and improved runner flow angles. However, overall improvement in performance may be disappointing unless the manifolds are installed as part of a complete performance upgrade.

For the TBI applications performance improvements intake manifold modifications may be more noticeable. The General Motors Performance Parts Catalog offers performance intake manifolds for use only with a two- or four-barrel carburetor. Holley sells a performance manifold for the 2.5 liter TBI engine.

Summary

Back in the 1960s, massive and very noticeable improvements in performance could be affected through relatively minor and often not very expensive modifications. In those days, the large brute engines such as those made by Chevrolet had tremendous unlocked potential. Although factory-equipped with about a 650cfm carburetor (depending on the year and exact application this varied), the Chevy 396 had a flow rate potential of 458cfm. Installing a carburetor with a greater flow rate would instantly unlock horsepower and torque hidden by the stock carburetor's inability to fill the manifold with air quickly enough when the throttle was matted.

In the 1970s, anything and everything was done to engines in order to maintain an acceptable performance level while remaining emission-legal. During the late 1970s and early 1980s, manufacturers once again began a quiet push toward more performance. This time, they did not have the luxury of simply bolting on a larger carburetor and dumping gallons of gas through the engine. As a result, today's intake systems are highly refined when they roll off the assembly line. Thus, extensive work on the manifold without considering other engine, computer, and sensor modifications may yield disappointing results. Again, it's hard to improve upon a good design.

Camshaft

The basic function of the camshaft is to open the intake valve as quickly, as smoothly, and as widely as possible. The cam leaves the valve open as long as possible to allow atmospheric pressure to push air and fuel into the combustion chamber. The camshaft must then begin (along with the assistance of the valve spring) to allow the intake valve to close in enough time for the combustion chamber to be sealed as the piston begins to move up on the compression stroke.

Actually, the intake valve remains open for a short time as the piston begins to move upward on the compression stroke. Leaving the intake valve open briefly allows the velocity of the air traveling through the intake system to continue to cram air into the cylinder even as the piston begins to move upward. This effect is even more pronounced on engines equipped with "high rise" intake manifolds such as on the Tuned Port Fuel Injection and Ford EEC-IV truck applications. This is why increasing the airflow potential beyond the capability of the engine's volumetric displacement will result in improved performance. Less restriction in the intake system will increase the cylinder charging capability.

The camshaft must also open the exhaust valve as far, as fast, and as smoothly as possible. The opening of the exhaust valve begins as the piston nears BDC (bottom dead center) on the crank power stroke. This ensures that the exhaust valve will be completely open when the piston begins to move upward on the exhaust stroke. The camshaft begins to close the exhaust valve near TDC (top dead center); however, the exhaust valve remains open for a short time and the piston begins its downward travel. This does two things: First, it takes advantage of the velocity of the exhaust gases exiting the combustion chamber to accelerate the incoming air-fuel charge. This increases the volumetric efficiency of the engine. The second reason for leaving the exhaust valve open as the intake valve opens is that as the incoming charge is accelerated, some of the outgoing exhaust gases will be slowed, causing them to remain in the combustion chamber. From the performance perspective, this is neither desirable nor efficient. However, it does produce an EGR (exhaust gas recirculation) effect, causing combustion temperature to be lower and reducing the output of oxides of nitrogen.

Several companies offer off-the-shelf camshafts for performance and economy. For obvious reasons, these performance cams are pretty much limited to the bigger engines, the 5.7, 5.0, and 4.3 liter, and the 60-degree V-6s. But, designing a performance cam that will effectively increase torque and horsepower for an electronic fuel-injected engine is not a job for mere mortals. Changes in lift, duration, centerline, and timing that have proved infallible in the past, today can cause power loss, underfueling, and overfueling. The reason is

There have been many changes in mufflers in the past 10-15 years. Changes to the extent that many racers are using them to improve performance.

that changing the camshaft will change intake manifold vacuum (pressure) and airflow rates. These changes will affect the readings of the MAP and MAF sensors. Changes in the MAP (or MAF) sensor readings at a given rpm or engine load may cause undesired changes in air-fuel ratio and ignition timing. Crane Cams, among others, has a line of camshafts designed specifically for use with electronic fuel injection. Because of the research and development that has gone into these cams it is unlikely that a custom-ground camshaft would work as well—unless it is ground by the best of the best.

If you decide to perform a camshaft transplant, be sure to check with your supplier about how it will affect the street legality of your car. At very least, you should consider that the increased valve overlap of a performance camshaft will tend to increase hydrocarbon emission at an idle. Since many states test emissions only at an idle, installation may lead to emission test failure. The California BAR considers camshaft replacements (other than with a stock grind) to be an unacceptable modification.

Exhaust System

Looking back again to the 1960s, we saw a fascination among street enthusiasts with bigger pipes and "freer" exhaust flow. Today's stock exhaust systems are a far cry from the stock systems of that decade. In spite of this, there is still room for improvement.

Improvement can begin with a set of headers bought over-the-counter, with threads and fittings for the oxygen sensor and the air pump devices. Catalytic converters with high flow rates are also available. Be sure to check with local automotive emission officials concerning the legality of any exhaust modifications you are planning to make.

Besides the catalytic converter, there is little that is not considered a legal modification in any state. Mufflers, connectors, and exhaust pipes are all fair game for the performance enthusiast. In most cases, the only restrictions are related to noise—which brings us to the subject of mufflers.

Mufflers

Exhaust system technology has come a long way since the muscle car days of the sixties. Today's high-performance mufflers exceed the flow potential of even open headers.

There are two things that travel down the piping from the exhaust manifold, exhaust gases and sonic vibrations, or frequencies. The movement of the frequencies through the exhaust system tends to pull the exhaust gases along in much the same way that ocean waves pull a surfer to shore. The effect of the exhaust gases being pulled through the exhaust system helps to scavenge the cylinder, improving the engine's breathing ability.

Several companies of the sixties marketed the Turbo muffler. This muffler was developed by Chevrolet to be used on the Corvair Turbo applications. It consisted of a hollow tube with fiberglass pressed against the sides of the tube to deaden sound. The problem was that as the sound vibrations (frequencies) entered the muffler, they were killed by the fiberglass packing, negating the surfer effect.

Today's high-tech performance mufflers are able to reduce sound without eliminating the surfer effect. Imagine, for a moment, that you are setting up a stereo in your living room. The only place you can find to put one of the speakers is in the center of the north wall. The only place you can find to put the other speaker is directly opposite the first, on the south wall, so the two speakers are facing each other squarely. The only place you can find to put your chair is halfway between the two speakers. When the speakers produce exactly the same sound, the frequencies being emitted will collide and cancel each other out, creating a dead zone. Modern high-performance mufflers take

Modern performance mufflers take advantage of resonances and harmonics to actually allow the exhaust system scavenge the exhaust gases and to an extent supercharge the intake. (Remember cross-flow and valve overlap?)

advantage of this phenomenon. As the frequencies and exhaust gases enter the muffler, they are divided and sent in two different directions only to be brought back together as they pass through the muffler. When they are brought back together, the identical frequencies that were split earlier collide and cancel each other out, like the speakers did. The end result is less sound without a loss of the surfer action.

Currently, mufflers such as those described here are being used on everything from road racers to sprint cars.

H-Pipes

Most exhaust systems have an area where the frequencies we have been discussing tend to build up and eddy the exhaust gases. To reduce this buildup on dual exhaust systems it helps to install an H-pipe between the two

Modern electronic fuel injection systems tempt enthusiasts with the dream of "plug-in" horsepower. In reality performance chips offer little to improve the overall performance of the engine. On the other hand, when camshaft and head modification have been done, the next step on these late model applications is to install a chip that has been programmed to take full advantage of the modifications.

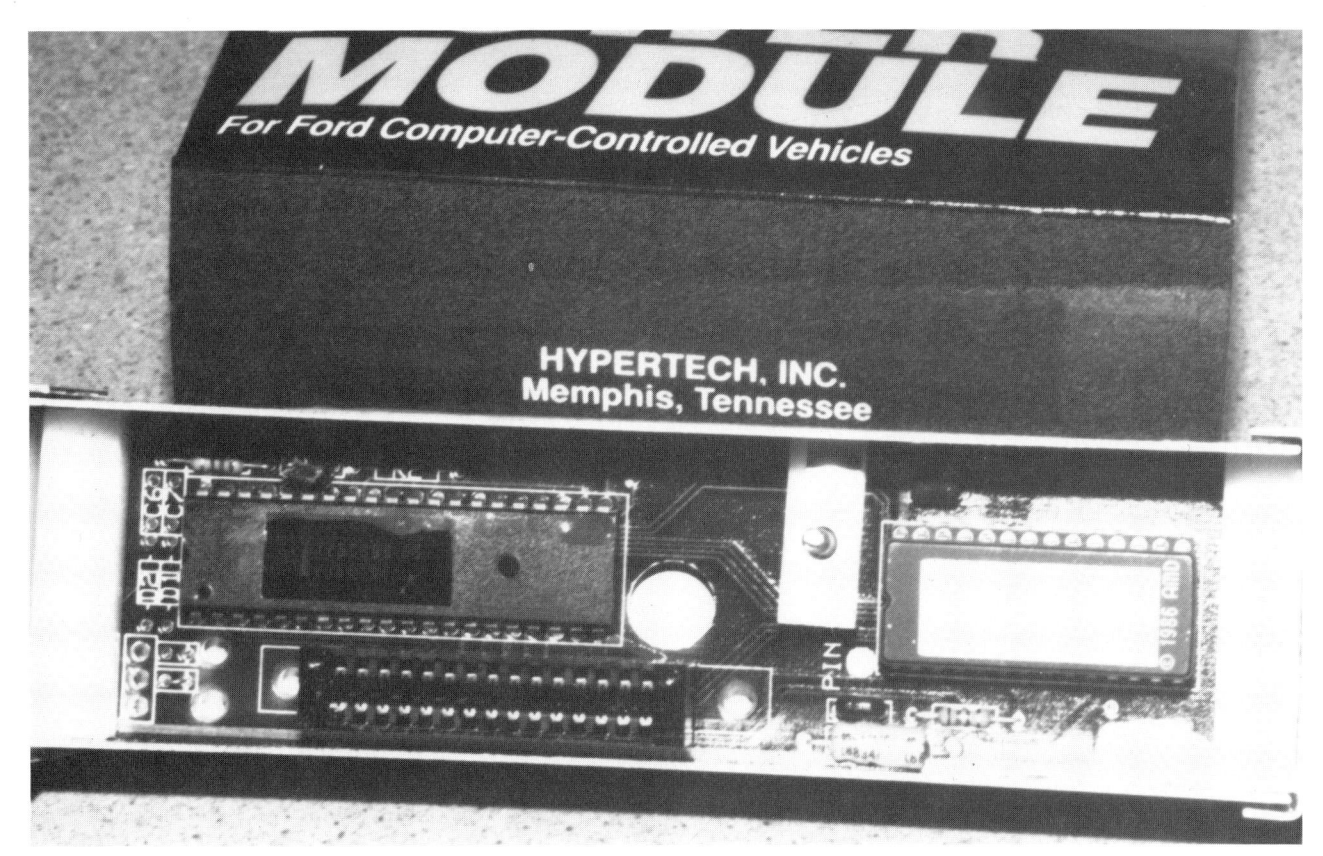

This chip for a Ford Mustang plugs into the end of the computer.

141

Horsepower from a thermostat? Now this is ridiculous, isn't it? Not really. By keeping the engine temperature lower than 180 degrees the computer allow the timing curve to advance more rapidly and allows for earlier enrichment of the fuel injection system under a load. The result? More horsepower.

sides of the exhaust. However, installing the H-pipe in the wrong place does more harm than not installing one at all.

To determine where an H-pipe is needed on your custom dual-pipe exhaust system, paint the area between the catalytic converters and the mufflers with black lacquer. Run the engine at 3200rpm for several minutes. Now inspect the painted area. Where the lacquer has begun to burn, or has burned extensively, is where the H-pipe needs to be installed. Install the pipe between the indicated hot-spots in the two sides of the exhaust.

Tuning the Engine

15

Once you've installed the cylinder heads and related equipment, if the engine fails to start, or does not run well once started, check the basics first. This is especially true if the engine is fuel injected or computer controlled. Computer-controlled engines have a tendency to divert even the best technicians from looking over the basic components. In the case where the engine will not start, or runs rough when started, check the fuel supply. Look into the carburetor as you move the throttle from the closed to wide-open position. The accelerator pump should spray fuel.

If the engine is fuel injected, check the fuel pressure. If the fuel pressure is okay, use a mechanic's stethoscope (heater hose held to the ear) to see if the injector is receiving pulses from the computer. Check the ignition timing. Remove the distributor to bring the number one piston to top dead center compression. Check the rotor and point or pickup coil alignment and reinstall. Check the firing order, making sure you begin numbering from the number one position on the distributor cap. Although you may be confident about the cam timing, remove one of the valve covers. Rotate the crankshaft until the rockers or cam lobe of the number one cylinder or its companion cylinder are at exactly overlap.

The timing marks on the crankshaft should indicate top dead center. If they do not, blame your helper and tear down the engine enough to reset the valve timing. If all else fails, use the more detailed troubleshooting information that follows.

Troubleshooting
Engine Will Not Start

No-start problems typically are the easiest to troubleshoot. There are three things required to get the engine started: fuel, air, and spark. Confirm that there is air and fuel available to the engine.

To check for air, connect a vacuum gauge to the intake manifold and crank the engine with the throttle closed. If the gauge reads 2 or 3in of vacuum or more, there should be enough vacuum to get the engine started. Failure to read vacuum could mean that the engine is cranking too slowly. Connect a tachometer to the negative terminal of the coil and crank the engine. If the cranking speed is less than 150rpm, recharge the battery and try again tomorrow. If the engine still cranks slowly, check for free crankshaft rotation. Does it turn freely by hand? If the engine rotates freely and the battery is charged, replace the starter.

If the engine cranking speed is adequate and yet there is still not enough vacuum, you must have set up the cam timing incorrectly. Confirm and correct.

Checking for fuel on carbureted applications is easy. With the ignition switch off, pump the accelerator several times while watching the carburetor venturis. Fuel from the accelerator pump should squirt down the venturis. If it does not, remove the fuel line from the carburetor and run a hose from the line into an approved fuel receptacle. Have someone crank the engine and observe the end of the line. If fuel pumps through, the carburetor needs repair. If fuel does not pump through, check for fuel line

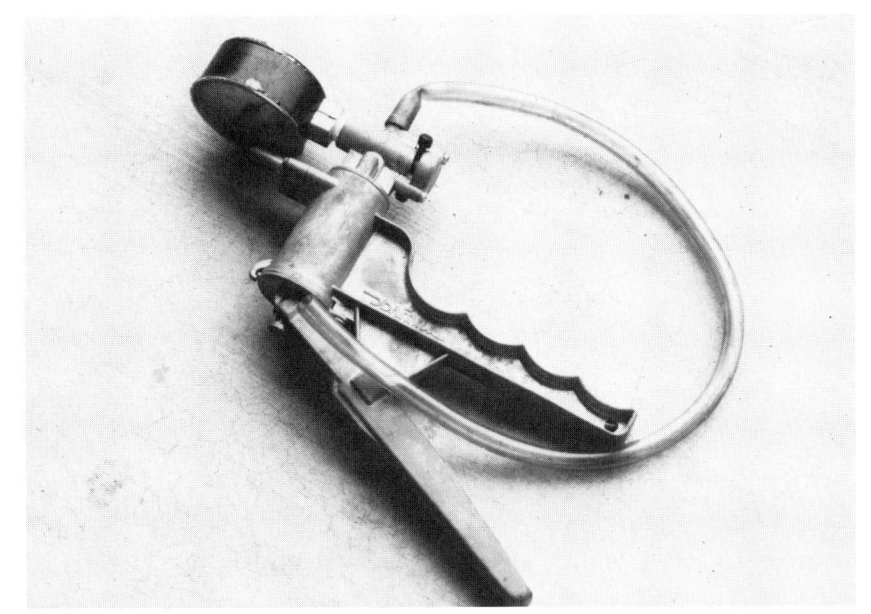

Vacuum tools like this are very handy for diagnosing problems. This one not only reads vacuum but can be used to create a vacuum when testing a vacuum controlled device or diaphragm

143

On a carbureted or throttle body injected engine fuel flow in a no-start condition can be detected visually. On multipoint fuel injected engines checking for fuel to the engine is a little more difficult. Install a fuel pressure gauge to confirm that fuel is available to the injectors. Use a mechanic's stethoscope to see if the injectors are clicking. If there is fuel pressure and the injectors are clicking, the no-start problem is not related to the injection system.

restrictions between the tank and the pump. If the fuel lines are in good condition, replace the fuel pump.

Checking for fuel delivery on throttle-body-injected engines is easy. While someone cranks the engine, observe the tip of the injectors; they should be spraying fuel into the throttle bore. If the injectors do not spray, use a mechanic's stethoscope to see if the injectors are clicking. If they are, then the vehicle is either out of fuel or the fuel pump is inoperative. Confirm that there is fuel in the tank. Check the fuses and power leads to the fuel pump with the engine cranking. (*Note:* There will be no power to the fuel pump of a fuel-injected car if the engine is not being cranked.) If the pump is receiving power and has a good ground, replace the fuel pump.

On multipoint fuel-injected engines, checking for fuel to the engine is a little more difficult. Install a fuel pressure gauge to confirm that fuel is available to the injectors. Confirm that there is fuel in the tank. Check the fuses and power leads to the fuel pump with the engine cranking. (*Note:* There will be no power to the fuel pump of a fuel-injected car if the engine is not being cranked.) If the pump is receiving power and has a good ground, replace the fuel pump. If the fuel pressure is about right (30-45psi for most multipoint applications), use a mechanic's stethoscope to see if the injectors are clicking. If there is fuel pressure and the injectors are clicking, the no-start problem is not related to the injection system.

To check for spark, disconnect one of the spark plug wires and perform the Ben Franklin test. Hand the plug wire to your nephew and crank the engine. If he begins to jump around, then you know you have good spark. Just kidding. In fact, if you like your nephew or fear reprisal or lawsuits from his parents, insert a screwdriver into the plug wire and hold the screwdriver about 1/4in from the engine block or cylinder head. Have someone crank the engine and observe the spark. If there is a good, crisp, blue spark, replace the spark plugs. If the spark does not fit this description, then move on to the next step.

Remove the coil wire from the distributor cap and hold it 1/4in from ground. Crank the engine. If you see a good, crisp, blue spark, replace the distributor cap and rotor. Please keep in mind there is a possibility that the problem is a defective set of plug wires. The reason I do not suggest their replacement at this point is that it is unlikely all of the plug wires went bad at the same time. If replacing the distributor cap and rotor does not cure the problem, replace the plug wires.

If there is not a good spark from the coil wire, connect a test light to the negative terminal of the coil. Crank the engine. If the test light blinks on and off as the engine is cranked, the problem is in the secondary side of the ignition system. Check the resistance of the coil wire. If the resistance is greater than 20,000 ohms, replace the coil wire. If the resistance is less than 20,000 ohms, replace the coil.

If the test light does not blink as the engine is cranked, the problem is in the primary. Move the test light to the positive terminal of the coil. Make sure the key is on. If the test light does not light, there is a problem in the power supply side of the primary ignition system. If the light does light, then the problem is in the distributor, ground side of the system.

If the problem is on the power side, crank the engine. If the test light illuminates while the engine is being cranked, check the resistance of the ballast resistor. The ballast resistor should have extremely low resistance. Replace the ballast resistor, if necessary. If the ballast resistor is good, repair the wiring in the bypass circuit. If the test light did not illuminate when the engine was cranked, the circuit from the ignition switch up to where the current flow divides to go through the coil and the ballast resistor.

If the problem is on the distributor side, remove the distributor cap and inspect the points. If they appear to be pitted or burnt, replace the points and condenser. Check again for a pulse at the negative terminal of the coil with the test light. If the test light pulses when the engine is cranked, then the engine should start. If the test light still does not pulse, check the distributor ground. If the distributor ground is good, repair the wire between the negative terminal of the coil and the distributor.

Engine Starts, but Does Not Continue to Run When Key Is Released

The most common cause of this classic point-condenser symptom is a defective ballast resistor. However, any open in the main 12-volt power supply to the ignition coil can cause this symptom.

Engine Misfires at Idle

Although an ignition misfire generally is in the secondary side of the ignition system, problems with the ignition points can cause symptoms that are exactly the same as defective spark plugs.

Before troubleshooting any misfire it is essential to verify that the engine is in good condition. A compression test is a good starting point. If the valves are adjustable, be sure they are properly adjusted.

With a pair of sissy pliers, remove and replace one plug wire at a time from the spark plugs. As each plug wire is removed, the engine rpm should drop. If one of the cylinders fails to produce as great a drop in rpm as the others, that cylinder is the source of the misfire.

Assuming the cylinder is in good condition and the valves are properly adjusted, remove the spark plug wire for that cylinder and check the resistance. The resistance should be less than 10,000 ohms per volt. If the resistance is correct, replace the spark plug. Unless the spark plugs are very new, replace them all.

Engine Misfires Under a Load

Assuming the engine is in good condition, begin troubleshooting this problem by checking the spark plug gap. If they

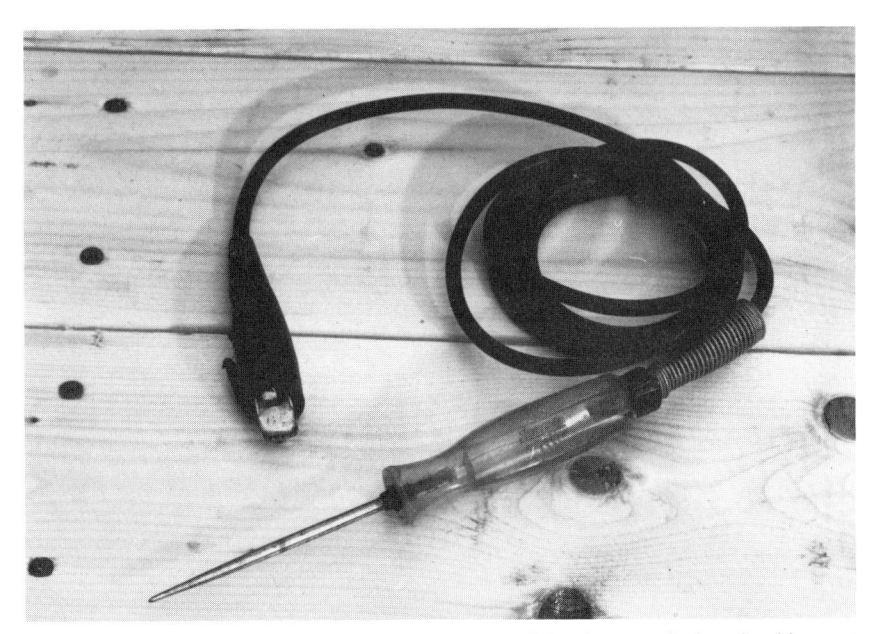

Even in this day of high tech computer controlled engines the test light is still one of the best tools for checking out the primary ignition system.

145

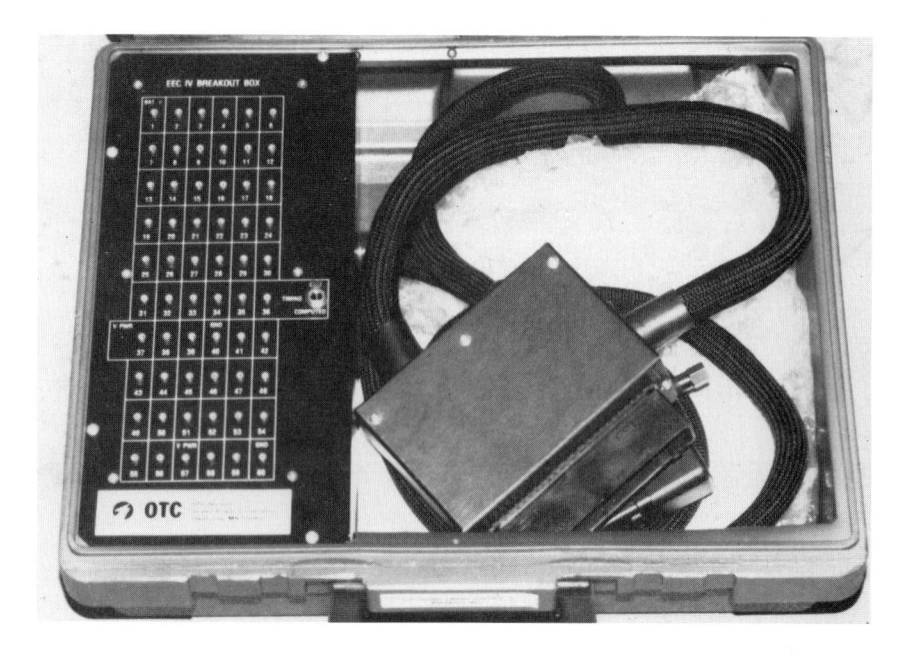

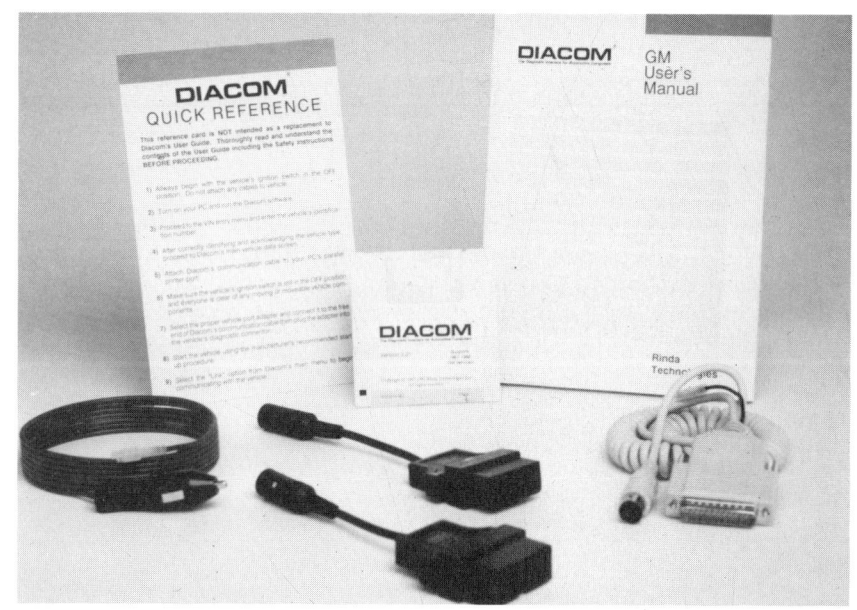

High-tech instruments are available for testing computer controlled fuel injection systems. However, if the problem you are looking for did not exist before the head work was done, then these instruments will probably not be necessary to find the cause of the problem now.

are gapped properly, replace the spark plugs. Even new spark plugs can misfire under a load.

If replacing the spark plugs does not solve the problem, remove the distributor cap. Inspect the wiring to the points. Frayed wiring can cause an intermittent open circuit as the vacuum advance moves the breaker plate. The intermittent open can cause a misfire.

Engine Lacks Power

There are many things that can cause a lack of power, some related to the ignition system, some not. Begin checking this problem by confirming that the engine and the air and fuel filters are in good condition.

If an ignition system problem results in a lack of power, it is likely the problem is in the timing control system. To test the timing control system, connect a timing light to the engine. Disconnect the vacuum advance and plug the hose. With the engine at idle speed, check the timing. Now raise the engine speed to 2000-2500rpm. If the timing does not advance, the centrifugal advance system is not working. Inspect the distributor weights. If they are free and move easily, replace the weight springs. If the springs are weak they will allow the timing to advance all the way prematurely, even at idle. If the weights are frozen, use penetrating oil or whatever is necessary to free them. If they are badly corroded it may be necessary to replace the distributor.

If, or when, the centrifugal advance is working properly, with the engine still at 2000-2500rpm, reconnect the vacuum hose to the vacuum advance. When the vacuum hose is reconnected, the timing should advance several degrees.

Testing the Sensors
Hall Effects Sensor

The Hall Effects sensor is often used as an alternative to the pickup coil. Many ignition systems, both distributorless and those with a distributor, use a Hall Effects device. Its primary advantage over the pickup coil is its ability to detect position and rotational speed from zero rpm to tens of thousands of rpm. Its primary disadvantage is that it is not as rugged as the pickup coil, and it is more sensitive to errant magnetic fields. An intense magnetic field can shut down the proper operation of a Hall Effects.

How does Hall Effects work? A Hall Effects pickup is a semiconductor carrying a current flow. When a magnetic field falls perpendicular to the direction of that current flow, part of that current

On a modern engine there are many little things that can cause in engine to run poorly. Many Ford applications, for example, use this device to control the position of the EGR valve. Inside this device is a filter. If the filter becomes restricted the EGR valve will be held open by vacuum and the engine will idle extremely rough.

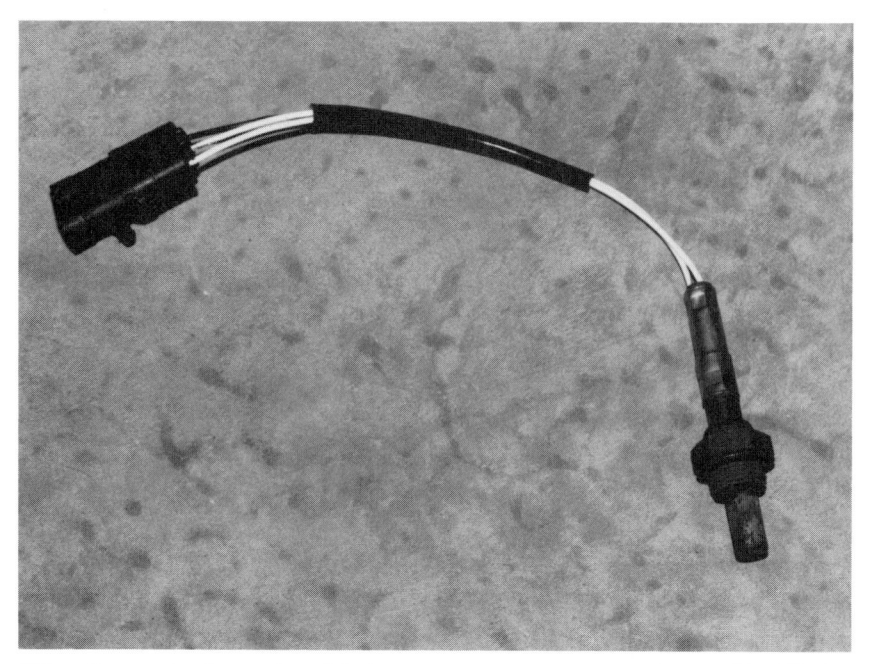

When the engine is running rich there is a low oxygen content in the exhaust, a high voltage (over 450 millivolts) is produced. When the engine is running lean there is a high percentage of oxygen in the exhaust and a low voltage (less than 450 millivolts) is generated. At the point of perfect combustion, known as the stoichiometric point, the oxygen sensor produces 450 millivolts. This is known as the crossover point. Many applications have a very high impedance circuit which will replace a missing oxygen sensor signal with a default voltage of 450 millivolts. This voltage can be detected with a high impedance voltmeter whenever the oxygen sensor is cold or disconnected.

Be careful selecting sealants for assembling the engine. If your engine is equipped with an oxygen sensor and if you decide to use a silicone sealer, be sure to use a "Low Volatile" or "oxygen sensor safe" silicone.

is redirected perpendicular to the main current path. The semiconductor is placed near a permanent magnet. A set of metal blades, or armature, attached to a rotating shaft or other device passes between the Hall Effects semiconductor and the permanent magnet. As the armature rotates, the magnet field is alternately applied to the Hall Effects and interrupted. The result is a pulsing current perpendicular to the main current path. This frequency is directly proportional to the speed of armature rotation. Since the output is dependent only on the presence of the magnetic field, the Hall Effects unit is capable of detecting armature position even when there is no rotational speed.

To test the Hall Effects sensor, connect a voltmeter to the output lead. The voltmeter should display either a digital high (four volts or more) or a digital low (around zero volts). Slowly rotate the armature while observing the voltmeter. If the voltmeter had read low, it should now read high; if the voltmeter had read high, it should now read low. If the voltage fluctuates in this manner as the armature is rotated, then the Hall Effects is good.

The Hall Effects can also be tested with a common dwell meter. Since the signal generated by the Hall Effects is a square wave, the dwell meter becomes a natural for testing. Connect the dwell meter between the Hall Effects output and ground. Rotate the armature as quickly as possible, for instance, by cranking the engine. The dwell meter should read something other than zero and full scale. If it does, the Hall Effects is good.

In addition to the dwell meter, the tachometer is also a good tool for detecting square waves. Connect the tachometer between the Hall Effects output and ground. With the armature rotating as described, the

There is nothing new under the sun. This is the MAP (Manifold Absolute Pressure) sensor from a Ford. It reports intake manifold pressure to the computer so the computer can control the functions that used to be controlled by the power valve and the vacuum advance. If the unit is defective or the vacuum lines are not correctly connected to it, severe driveability problems can result.

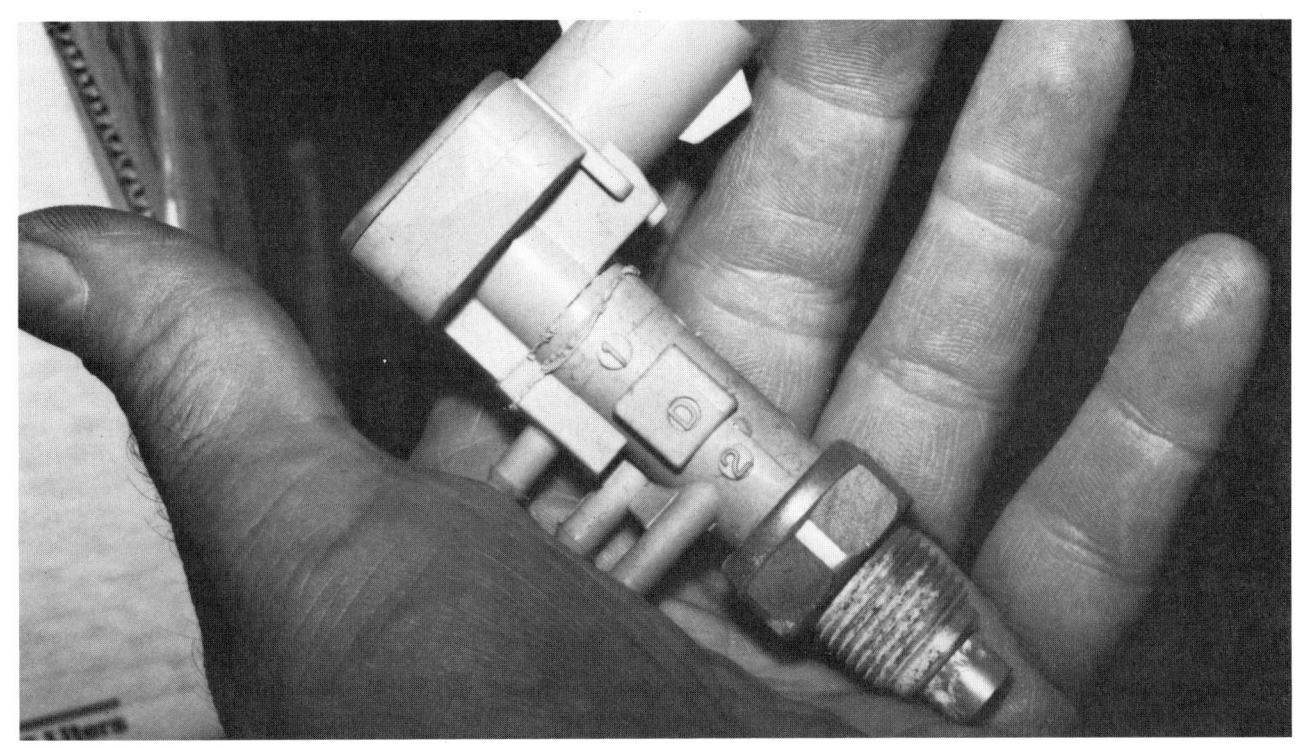

In the "good ol' days" a misrouted vacuum line had little or no affect on the operation of the engine. Do not throw these devices away just to make it look neater under the hood.

Many ignition systems, both distributorless and distributor type, use a Hall Effects device. These devices are relatively fragile and should be replaced when major engine work is done or every 60,000 to 80,000 miles.

tachometer should read something other than zero if the Hall Effects is good.

Optical Sensor

The optical sensor is used on some Nissan products and the 3.0 liter Mitsubishi Chrysler engine. The signal produced by the optical sensor is identical to the one produced by the Hall Effects. However, the signal is produced by an armature interrupting light. A light-emitting diode or LED (usually emitting infrared, invisible light) sits opposite an optical receiving device such as a photodiode or phototransistor. An armature is rotated between the LED and the receiver. Unlike the Hall Effects, the armature in the optical sensor can be made of metal, plastic, or any translucent material. As the armature rotates, light alternately falls on and is kept from falling on the receiver. As this occurs, the current flowing through the receiver is turned on and off creating a square wave with a frequency directly proportional to armature rotation. In some cases, including many GM vehicle speed sensors, the light is reflected off of rotating blades rather than being interrupted.

The main advantage of the optical rotational sensor over the pickup coil and the Hall Effects is its ability to produce extremely high frequencies. The 3.0 liter Chrysler distributor produces a frequency of 540,000 hertz (0.54 megahertz!) at just 3000rpm. The primary disadvantage is its sensitivity to dirt, oil, and grease, which will create erroneous signals.

The optical sensor is tested in the same ways recommended for testing the Hall Effects sensor.

Testing and Replacing the Distributor Cap

Testing the distributor cap is done visually. Look for cracks and carbon tracks. Even the most trained eye often will miss these problems, so perhaps the best way to check the distributor cap is to replace it. Replacement consists of simply releasing the attachment screws or clips and making sure that the plug wires are installed in the correct order.

Whichever type of distributor you have on your car, it would be a good idea to replace the distributor cap and rotor together and use the same brand. Pairing caps and rotors of two different manufacturers is not a good idea as this can result in the incorrect rotor air gap. Excessive rotor air gap can cause excessively high spark initiation voltage, and can result in incomplete combustion.

Replacing the Air Filter

The real value of the air filter is significantly underestimated. It is the engine's only defense against sand, grit, and other hard particle contamination. When these substances enter the combustion chamber they can act like grinding compound on the cylinder walls, and piston rings and valves. Replace the air filter at least once a year or every 24,000 miles. In areas where sand blows around a lot, like west Texas or Arizona, the air filter should be replaced much more often.

On most carbureted cars a restricted air filter will cause the engine to run rich. This is because the restriction causes a reduction of pressure in the carburetor venturi while the pressure in the fuel bowl remains constant

---

A Hall Effects pickup is a semiconductor carrying a current flow. When a magnetic field falls perpendicular to the direction of that current flow part of that current is redirected perpendicular to the main current path. The semiconductor is placed near a permanent magnet. A set of metal blades, or armature, attached to a rotating shaft or other device passes between the Hall Effects semiconductor and the permanent magnet. As the armature rotates the magnet field is alternately applied to the Hall Effects and interrupted. The result is a pulsing current perpendicular to the main current path. This frequency is directly proportional to the speed of armature rotation. Since the output is only dependent on the presence of the magnetic field, the Hall Effects unit is capable of detecting armature position even when there is no rotational speed.

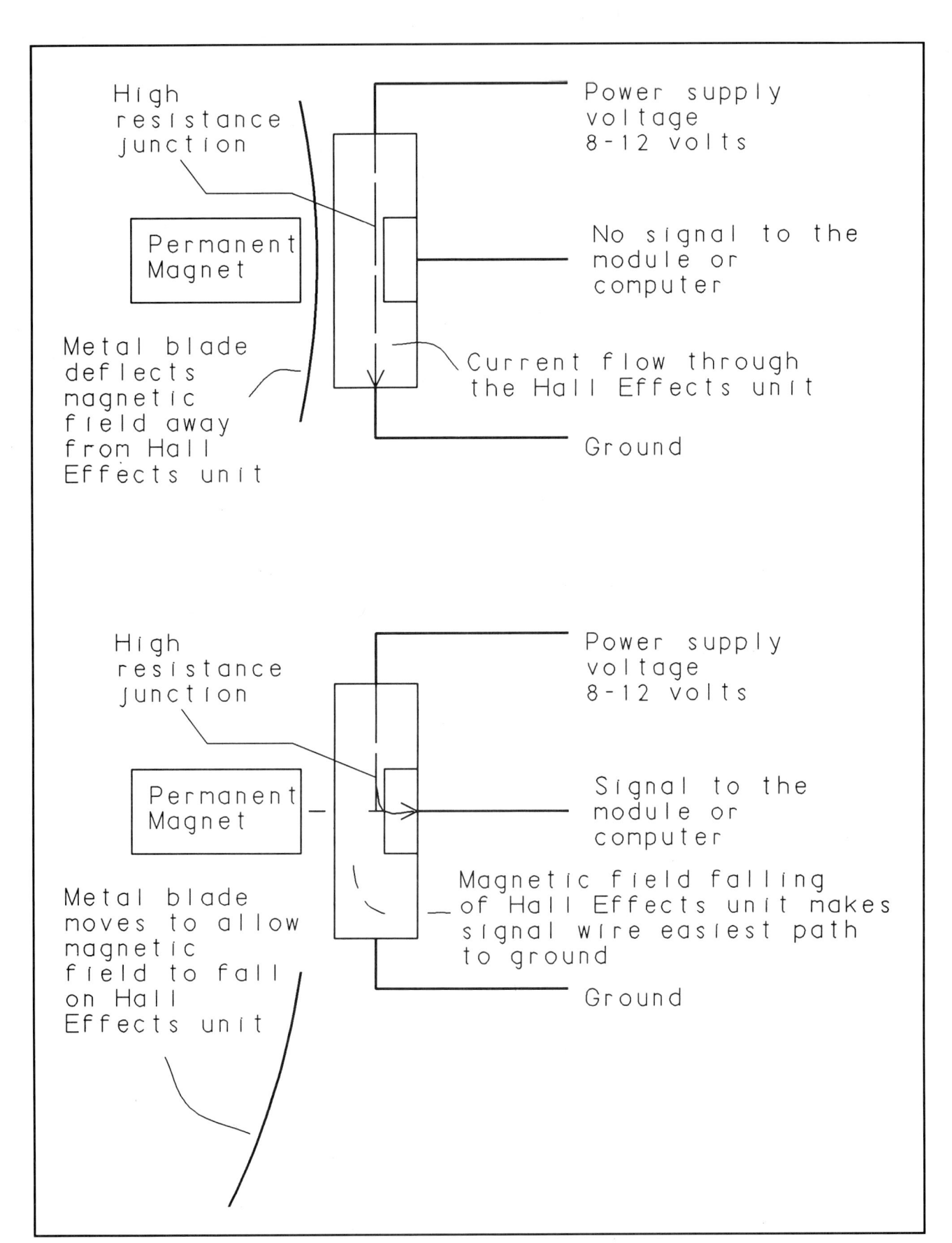

Top diagram:

- High resistance junction
- Permanent Magnet
- Metal blade deflects magnetic field away from Hall Effects unit
- Power supply voltage 8-12 volts
- No signal to the module or computer
- Current flow through the Hall Effects unit
- Ground

Bottom diagram:

- High resistance junction
- Permanent Magnet
- Metal blade moves to allow magnetic field to fall on Hall Effects unit
- Power supply voltage 8-12 volts
- Signal to the module or computer
- Magnetic field falling of Hall Effects unit makes signal wire easiest path to ground
- Ground

Infra-red photo-diode

Infra-red light emitting diode

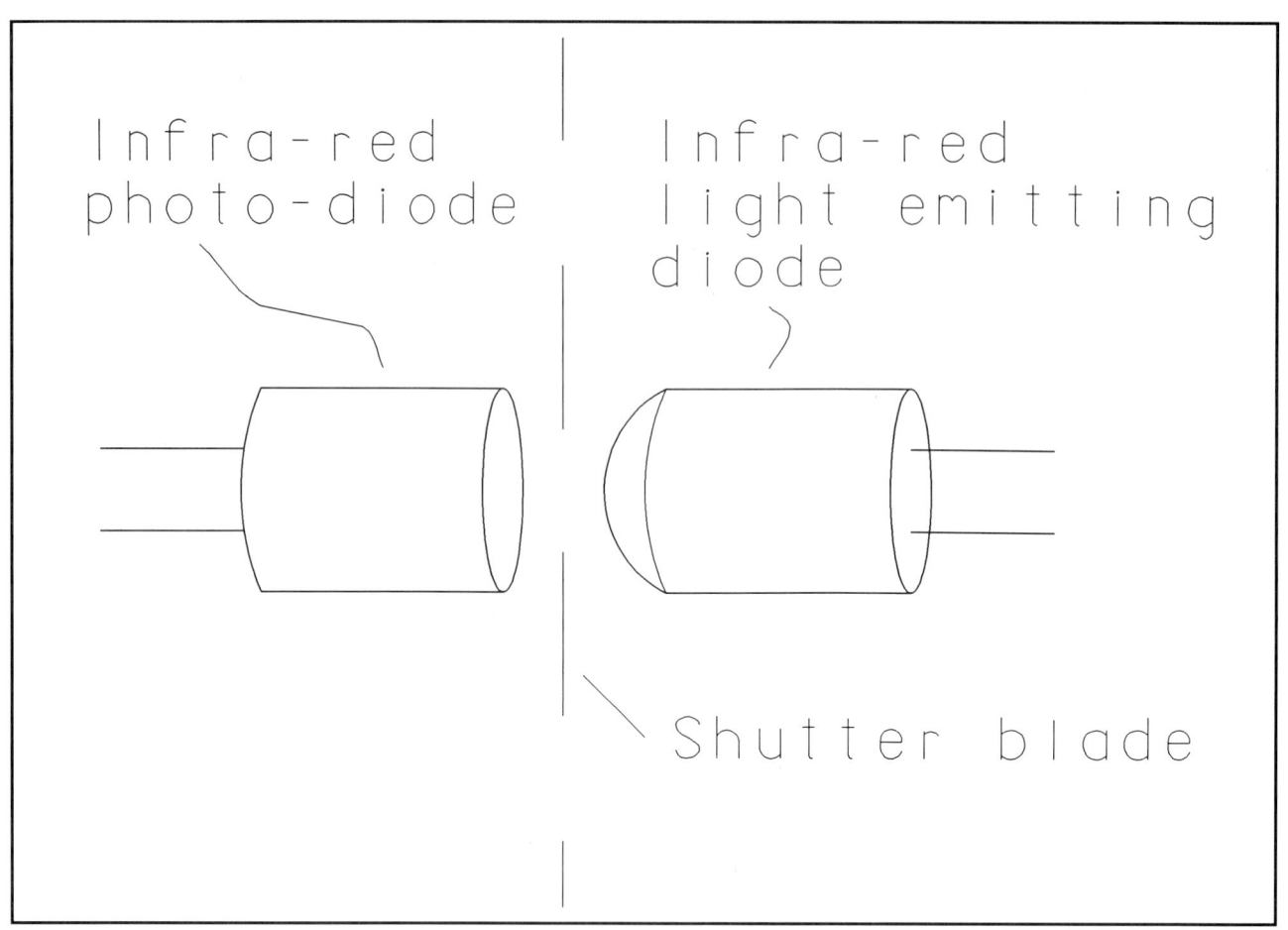

Shutter blade

The signal produced by the optical sensor is identical to the one produced by the Hall Effects. The signal, however is produced by an armature interrupting light. A light emitting diode (usually infra-red, invisible light) sits opposite an optical receiving device such as a photo diode or photo transistor. An armature is rotated between the LED and the receiver. Unlike the Hall Effects, the armature in the optical sensor can be metal, plastic or any translucent material. As the armature rotates, light alternately fall on and is kept from falling on the receiver. As this occurs the current flowing through the receiver is turned on and off, creating a square wave with a frequency directly proportional to armature rotation. Among the applications to use this sensor as a distributor pickup is the 3.0 Dodge engine.

at atmospheric. This increased pressure differential increases the flow of fuel into the venturi and the mixture enriches. On a fuel-injected engine, the ECU measures the exact volume of air entering the engine with the airflow meter, and delivers the correct amount of fuel for the measured amount of air. Restrict the flow of incoming air and less air will be measured, less fuel will be metered into the engine. In applications using a MAP sensor, air measurement is not very precise and therefore may run a little rich as a result of a restricted air filter.

Replacing the Fuel Filter

The fuel filter is the most important service item among the fuel components of the fuel injection system. In my twenty years of experience with fuel-injected cars, I have replaced many original fuel filters on cars that were over ten years old. This lack of routine maintenance is just begging for trouble.

After removing the fuel filter, find a white ceramic container, such as an old coffee cup, and drain the contents of the filter through the inlet fitting. Inspect the gasoline in the cup for evidence of sand, rust, or other hard particle contamination. Now pour the gas into a clear container such as an old glass and allow it to sit for about thirty minutes. If there is a high water content in the fuel, it will separate while sitting. The fuel will float on top of the water. If there is excessive water or hard particle contamination in the tank it may have to be removed and professionally cleaned. For minor water contamination problems there are additives that can be purchased at your local parts house.

Should the fuel filter become excessively clogged, the following symptom might develop. You start the car in the morning and it runs fine. As you drive several miles down the road, the car may begin to buck a little or lose a little power, then suddenly the engine quits almost as though some one had shut off the key. After sitting on the side of the road for several minutes, the car can be restarted and driven for a couple of miles before the symptom recurs. This could be caused by a severely restricted fuel filter. The car runs well initially because the bulk of what is causing the restriction has fallen to the bottom of the fuel filter as sediment. When the engine is started and the fuel begins to flow through the injection system, this sediment gets stirred up and presses against the paper elements of the fuel filter. As this occurs, fuel volume to the injectors is decreased and the engine begins to run lean. Sooner or later the engine leans out so much that it dies. The really bad news is when the fuel filter becomes that restricted, some of the contaminants have forced their way through the filter and may contaminate the rest of the fuel system.

Testing for Alcohol Contamination

Alcohol contamination can damage many of the fuel injection components. Use of gasohol can be one source of alcohol. If you suspect that excessive alcohol content has caused the failure of system components, then you might want to test a fuel sample for alcohol content. Pour 200 milliliters of the sample fuel into a glass or clear plastic container, along with 100 milliliters of water. Immediately after putting the two liquids in the container the dividing line will be at the 100 milliliter mark. Wait about thirty minutes. If the dividing line rises by more than 10 per-

I once saw a transmission rebuild because of a dirty fuel filter. If you have gone to the trouble and expense of doing performance modification on you heads, does it not also make sense to go to the relatively small expense of replacing the fuel filter?

cent of the volume of the contents of the container, then there is excessive alcohol in the fuel. Drain the fuel from the tank and replace it with good fuel.

There are many fuel additives on the market that contain alcohol. However, the additive

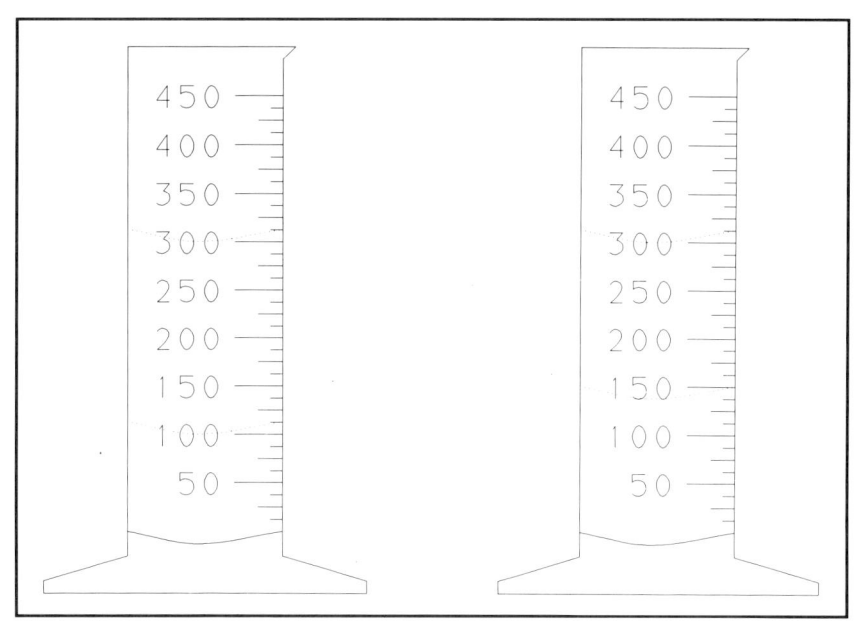

Pour 200 milliliters of the sample fuel into a glass or clear plastic container along with 100 milliliters of water. Immediately after putting the two liquids in the container the dividing line will be at the 100 milliliter mark. Wait about thirty minutes, and if the dividing line rises by more than 10 percent of the volume of the contents of the container, then there is excessive alcohol in the fuel. Drain the fuel from the tank and replace it with good fuel.

153

Whether the engine you are working on is a late model Cadillac Northstar engine or an old Model A, the rules of troubleshooting are still the same.

content is so small compared to the size of the typical fuel tank that they pose no threat to the fuel system. Nevertheless, use caution, ask around a bit, and be selective when purchasing these products. Some are much better than others.

Testing for the Dead Hole

When a late-model fuel-injected engine has a cylinder with a misfire, the effects can go far beyond a rough idle or loss of power. A cylinder that is still pulling in air but not burning that air will be pumping unburned oxygen past the Lambda sensor. This confuses the ECU, making it believe that the engine is running lean. The ECU responds by enriching the mixture, and the gas mileage deteriorates dramatically.

There are several effective methods that can be used to isolate a dead cylinder. All of these methods measure the power produced in each cylinder by killing them one at a time with the engine running a little above curb idle.

Back in the good old days, we used to take a test light, ground the alligator clip, and pierce through the insulation boot at the distributor cap end of the plug wire. This would ground out the spark for one cylinder and an rpm drop would be noted. The greater the rpm drop, the more power that cylinder was contributing to operation of the engine. Actually, this is a valid test procedure; however, piercing the insulation boot is only asking for more problems than your started with.

Another "old days" method of performing a cylinder balance was to isolate the dead hole by pulling off one plug wire at a time and noting the rpm drops. The problem with this method is that you run the risk of damaging either yourself or the ignition module with a high-voltage spark.

So, let's explore some alternatives.

Testing for Weak Cylinders

Several tool companies produce a cylinder-shorting tach or dwell meter. These devices electronically disable one cylinder at a time while displaying rpm. Engine speed drops can be noted. Unfortunately, these testers can cost $500 or more.

Another method does the old test light technique one better. Cut a piece of 1/8in thick vacuum hose into four, six, or eight sections each about 1in long. With the engine shut off and one at a time, so as not to confuse the firing order, remove a plug wire from the distributor cap, insert a segment into the plug wire tower of the cap, and set the plug wire back on top of the hose. When you have installed all the segments, start the engine. Touching the vacuum hose conductors with a grounded test light will kill the cylinder so that you can note rpm drop. Again, the one with the smallest drop in rpm is the weakest cylinder.

Whichever of the above methods you use, follow this procedure for the best results:

1. Adjust the engine speed to 1200-1400rpm by blocking the throttle open. Do not hold the throttle open by hand, you will not be steady enough.
2. Electrically disconnect the idle stabilizer motor to prevent it from affecting the idle speed.
3. Disconnect the Lambda sensor to prevent it from altering the air-fuel ratio to compensate for the dead cylinder.
4. Perform the cylinder kill test. The rpm drop should be fairly equal between cylinders. Any cylinder that has a considerably smaller rpm drop than the rest is weak. Proceed to step 5.
5. Introduce a little propane into the intake, just enough to provide the highest rpm. Repeat the cylinder kill test. If the rpm drop from the weak cylinder tends to equalize with the rest, then you have a vacuum leak to track down. If the car has an idle stabilizer, be sure to disable it so that it will not attempt to compensate for the lack of power input from the dead cylinder.

Point-Condenser

The point-condenser ignition system served as the workhorse ignition system of the gasoline engine for more than fifty years. Components of the primary circuit include the following:

Coil

The heart of any spark ignition system is the coil. Basically, the ignition coil used in the point-condenser ignition system is a step-up transformer inside of an oil-filled metal can. As there are many times more secondary windings as primary windings, the voltage output of the secondary is potentially tens of thousands of volts.

Although the coil is relatively trouble free, both the primary and secondary windings are subject to opens, shorts, grounds, and corrosion. When I was in trade school, and I confess it was in the days when electronic ignition was viewed as a gimmick and aberration, my tune-up instructor pointed out the valid tell-tale sign of a coil defect. Most point-condenser ignitions have a recess in the bottom similar to an aluminum soda can. An internal problem in the coil can cause overheating which can cause the oil intended to cool the coil to expand, pushing out the recessed area at the bottom of the coil. Although this evidence is not conclusive as to coil condition, it can serve as one piece of the diagnostic puzzle.

Ignition Switch

The ignition switch is normally a key lock switch that serves as an on-off switch for primary current flow. In the typical point-condenser ignition system there are two circuit paths through the switch for the primary side of the coil. The first path carries current to the ignition coil through a ballast resistor that reduces the voltage drop across the coil primary to about 9.6 volts, and limits current flow to about 2.5 amps. The sec-

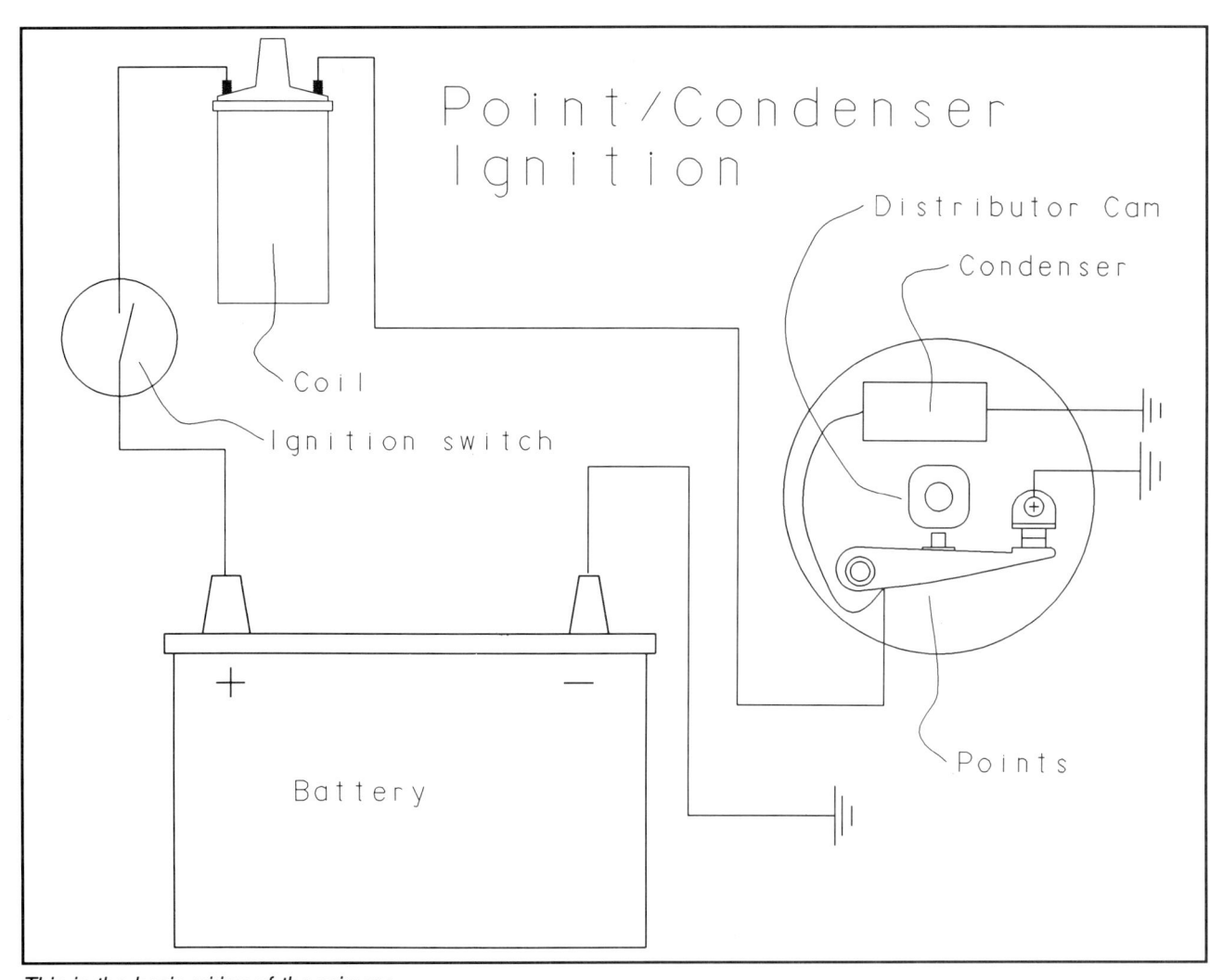

This is the basic wiring of the primary circuit of a point/condenser ignition system.

ond path carries current directly to the ignition coil when the engine is being cranked for starting. Bypassing the ballast resistor increases current flow and the available secondary power while starting the engine. This is especially important when the engine is being cold-started.

The ignition switch is one component of the ignition system that is often overlooked during the troubleshooting procedure. Although seldom the cause of driveability problems, an intermittent open circuit in the ignition switch can cause stalling and other power and performance related problems.

Battery

Most people, including most professional automotive technicians, do not think of the battery as part of the ignition system. Yet, the battery is the ultimate power source for every electrical and electronic component on the car. A intermittent open circuit in the battery can manifest itself in a wide variety of ways, including intermittent coil operation. Intermittent coil operation can cause stalling and misfiring.

Ballast Resistor

The ballast resistor is a low-resistance, high-wattage resistor that is installed in the primary circuit to limit current flow. In addition to limiting current flow, it also reduces the voltage drop across the ignition coil primary. This limiting of current and voltage reduces the amount of wear on the points by lowering their tendency to arc as they are opened to interrupt primary current flow.

The ballast resistor is located in the run circuit between the ignition switch and the coil. The bypass position of the ignition switch routes coil current around the ballast resistor while the engine is being cranked. This provides full coil power to the spark plugs while starting the engine, particularly important when trying to cold-start the engine.

If all this potential power can be created with full current flow to the ignition coil, why even have a ballast resistor? If the ballast resistor is eliminated from the system, then arcing and excessive wear can occur at the points.

Condenser

The ignition condenser is an electrolytic capacitor. The device consists of a small metal can with two thin foil strips inside, separated by a thin insulating material. The condenser acts like an electrical shock absorber to the primary ignition system. When the points open, the primary current attempts to keep flowing. Without the condenser, the current would continue to flow across the opening points. This arcing would slow the collapse of the magnetic field and limit the potential power of the secondary ignition.

The condenser is seldom the cause of driveability problems. In fact, during the heyday of the point-condenser system, I worked for several repair shops that routinely did not replace the condenser during a tune-up. Yet a defective condenser can cause poor idling, no-start, and misfiring.

Points

Points are little more than a cam-operated switch. It is the opening and closing of the points that controls the current flow through the coil. As the crankshaft rotates, it drives the camshaft through gears or chains. A gear on the camshaft drives the shaft of the distributor. As the distributor shaft rotates, a cam is turned by the shaft. This rotating cam opens and closes the points in a synchronized fashion to the rotation of the crankshaft and the movement of the pistons.

Points are an inadequate method of controlling current for several reasons. First, as the points open there is a tendency for the current to continue flowing. This tendency can cause an arc and slow the speed of collapse of the magnetic field around the primary windings of the coil. As the speed of collapse slows, the potential output of the coil is decreased. In order to control this tendency to arc, current flow through the primary is kept to a minimum. This also limits the potential output of the system. A second factor is that the points become pitted and worn over time, increasing the resistance across the points and limiting the current flow. The rubbing block can wear as well, which will alter the timing. Finally, at extremely high engine rpm the points will tend to bounce open when they close.

Several changes have been made over the years to help eliminate these problems. Special metals were used at the contact points to decrease the wear and resistance factors. Holes were drilled in the contact points to reduce the possibility of arcing. But most of these modifications only delayed the inevitable.

Since pitting and wear of the points can decrease the potential current flow through the primary, and since current flow in the primary can affect the output of the secondary, points need routine replacement. If the condenser is not perfectly matched to the rest of the primary ignition system, point pitting can be accelerated. Pitted points can cause erratic coil operation which will result in misfire, stalling, and difficulty in starting. Thus, ignition points need to be replaced at least every 15,000 miles to ensure dependability of the ignition system.

Ignition Timing Control

Ignition timing must change as the engine is running to adjust to different engine speeds and loads. When the engine is running at idle, the spark must begin at a point in crankshaft rotation that will allow for the spark to extinguish when the crankshaft is about 10 degrees after top dead center. Since the length of time the spark is jumping the gap is a relative constant, about 2.5 milliseconds, the spark must start sooner as the engine speed increases.

Centrifugal Advance

On a hypothetical engine, the spark occurs at 10 degrees before TDC (top dead center) when the engine is running at 1000rpm. At this speed the spark extinguishes at about 10 degrees after TDC. This means that the crankshaft has rotated 20 degrees since spark initiation. As engine speed increases, the crankshaft rotates more degrees in the 2.5 milliseconds that the spark is jumping the gap. At 2000rpm the crankshaft will rotate twice as much. If the timing at 1000rpm should be 10 degrees before TDC, then the timing at 2000rpm should be about 30 degrees before TDC. As the engine speed continues to increase, the timing will need to continue to advance. The amount of total advance, the upper limit of the advance, will vary depending on the design of the engine.

The change of timing in response to rpm is accomplished through a set of spring-loaded weights. As engine speed increases, the weights swing out against spring tension. The cam that opens and closes the points, although mounted on the distributor shaft, is not part of the distributor shaft. The swinging weights cause the cam to rotate with respect to the distributor shaft. This advances the timing.

Vacuum Advance

At first glance, this is a misnomer. Vacuum advance actually *retards* the timing when the engine is under a load. In most applications, vacuum advance is connected to ported vacuum. The advance unit receives no vacuum at idle, but when the throttle is opened the vacuum advances the timing. As the load on the engine increases, the vacuum drops. As vacuum drops, the timing is not advanced as much, it retards. Retarding the timing lowers the combustion temperature, therefore preventing detonation and decreasing potential damage to the engine.

Setting the Timing

After the engine is running reasonably well, it is time to set the ignition timing. Keep in mind that if the heads and valves have been modified, if the compression ratio has been changed, or if the camshaft has been replaced, the timing requirements will be different than stock. There are almost as many different procedures to follow when adjusting initial timing in point-condenser ignition systems as there are different applications of point-condenser systems. The basic premise of adjusting or checking initial timing is to disable the timing controls. Most point-condenser systems have a centrifugal advance system which advances the timing as the speed of the engine increases. A second timing control system is the vacuum advance, which actually retards the timing when the engine is under a load.

For most applications, disconnect and plug the vacuum advance hose and reduce the engine idle speed as much as possible. Under these conditions any vacuum that might be in the hose cannot affect the timing, and the engine speed will be too low for the centrifugal weight to advance the timing.

This procedure will usually work when the correct procedure for your application is not available. Be sure to follow the exact procedure for the application you are working on. The information can be found on the decal under the hood.

If the vehicle is one of the late-model applications that have computerized control of ignition timing and the computer fails to control ignition timing properly, refer to these Motorbooks International publications as appropriate: *How to Repair and Modify Chevrolet Fuel Injection, How to Tune and Modify Ford Fuel Injection,* or *How to Tune and Modify Bosch Fuel Injection.* If these books do not cover the application you are working on, consult the factory service manuals.

If the camshaft has been replaced with a performance cam, use the recommendations made by the camshaft manufacturer. If the camshaft has been custom ground or if for some reason the camshaft manufacturer cannot give you a recommended timing setting, set the timing at the original, stock position plus 5 degrees of advance. If the engine performs well, advance the timing another 5 degrees. If the engine runs well again, advance the timing again. When the performance drops off, or when the engine begins to ping or detonate, retard the timing to the previous setting.

The Law and Engine Modifications 16

Unlike most scientific regimens the law has subtle twists, turns, and traps. This chapter is not intended to serve as legal advice, rather it is intended to point out some of the issues involved in high-performance modification of late-model cars. The bulk of the information came from California's Bureau of Auto Repair (BAR) and the California Air Research Board (CARB). I found that many state laws are similar to those of California; however, since California has the longest record of stringent enforcement, I will use their information in this chapter.

There are three categories of replacement parts recognized by CARB:

Category 1

Category 1 items are not considered by BAR or CARB to be of any concern as long as the required emission controls are not tampered with.
- PCV air bleeds
- Air cleaner modification
- Air conditioner cut-out systems
- Anti-theft systems
- Blow-by oil separators and filters
- Electronic ignition systems retrofitted to vehicles originally fitted with point-condenser systems as long as the original advance controls are maintained
- Engine shutoff systems
- Ignition bridges and coil modifications
- Throttle lockout systems
- Intercoolers for OEM turbocharger
- Under-carburetor screens
- Vapor, steam, and water injectors

Category 2

Category 2 addresses allowable replacement parts:
- Headers on noncatalyst cars
- Heat stoves for allowed headers
- Intake manifolds for non-EGR vehicles must allow for the installation and proper functioning of OEM emission controls
- Approved aftermarket catalytic converters
- Carburetors marketed as "emission replacement"
- Replacement fuel fill-pipe restrictors
- Replacement gas caps

As you can see from this list that for catalyst-equipped fuel-injected cars, no performance replacements are allowable without type approval from CARB. Today's cars are EPA inspected as an integrated system; disturbing even the minutest portion of the emission control package would constitute a violation. Category 3 parts must have verification of acceptability. If you are replacing a part in Category 3, ask for and retain a copy of the verification of acceptability for that product. It may prove handy later on, even if you live in an area that is not strictly controlled.

Category 3

Category 3 includes:
- Carburetor conversions
- Carburetors that replace OEM fuel injection
- EGR system modifications
- Replacement PROMs (computer chips)
- Electronic ignition enhancements for computerized vehicles
- Exhaust headers for catalyst vehicles
- Fuel injection systems that replace OEM carburetors
- Superchargers
- Turbochargers

Although not specifically stated in the California Air Resources Board rules or in the rules of the California Bureau of Auto Repair, cylinder head modifications or the computer modifications required to take advantage of the head modifications may violate the law.

Drill Sizes for Taps

Nominal drill sizes for taps, American threads

Nominal size	Tap drill, inches	Tap drill, mm
1/16-64	3/64	1.2
3/32-48	#49	1.85
1/8-40	#38	2.6
5/32-32	1/8	3.2
5/32-36	#30	3.25
3/16-24	#26	3.75
3/16-32	5/32	4
7/32-24	#16	4.5
7/32-32	3/16	4.8
1/4-20	13/64	5.1
1/4-24	#4	5.3
1/4-28	7/32	5.5
1/4-32	7/32	5.6
5/16-18	F	6.5
5/16-24	I	6.9
5/16-32	9/32	7.2
3/8-16	5/16	8
3/8-24	Q	8.5
7/16-14	3/8	9.3
7/16-20	25/64	9.9
1/2-12	27/64	10.75
1/2-13	27/64	10.75
1/2-20	29/64	11.5
9/16-12	31/64	12
9/16-18	33/64	13
5/8-11	17/32	13.5
5/8-18	37/64	14.5
11/16-11	19/32	15
11/16-16	5/8	16
3/4-10	21/32	16.5
3/4-16	11/16	17.5

Nominal drill sizes for taps, metric

Nominal size	Pitch, m/m	Tap drill, m/m
1.5	0.35	1.1
2	0.40	1.6
2	0.45	1.5
2	0.50	1.5
2.3	0.40	1.9
2.5	0.45	2
2.6	0.45	2.1
3	0.50	2.5
3	0.60	2.4
3	0.75	2.25
3.5	0.60	2.9
4	0.70	3.3
4	0.75	3.25
4.5	0.75	3.75
5	0.75	4.25
5	0.80	4.2
5	0.90	4.1
5	1.00	4
5.5	0.75	4.75
5.5	0.90	4.6
6	1.00	5
6	1.25	4.8
7	1.00	6
7	1.25	5.8
8	1.00	7
8	1.25	6.8
9	1.00	8
9	1.25	7.8
10	1.25	8.8
10	1.5	8.6
11	1.5	9.6
12	1.25	11
12	1.50	10.5
12	1.75	10.5
13	1.50	11.5
13	1.75	11.5
13	2.00	11
14	1.25	13
14	1.75	12.5
14	2.00	12
15	1.75	13.5
15	2.00	13
16	2.00	14
17	2.00	15
18	1.5	16.5
18	2.00	16
18	2.50	15.5
19	2.50	16.5
20	2.00	18
20	2.50	17.5
22	2.50	19.5
24	3.00	21
26	3.00	23
27	3.00	24
28	3.00	25
30	3.50	26.5

Index

Air cleaners, 135-136
Air filters, 150-152
Air-fuel mixture, 15-16
Alcohol contamination, 153-154

Bernoulli's Law, 11-12
Blueprinting, 93-117
Bolts, 28-30
    repairing and replacing, 32-34
    torquing, 28-29
Boyle's Law, 9-11
Bureau of Auto Repair (BAR), 158

California Air Research Board (CARB), 158
Cam bearing journals, 46, 74
Camshafts, 48-55, 70-71, 139-140
Camshafts, overhead, 94, 129, 131
CC gauges, 26-27
Chemical block tester, 25
Combustion chambers, 16-18
Compression tester, 19
Cylinder head
    measuring, 68-69, 73-74, 77-82, 95-96
    milling, 120
    polishing, 117
    porting, 110-116
    reassembly, 126-129
    repairing, 62-67
    resurfacing, 38-42
    selection, 35-36
    testing, 36-37, 63-64
Cylinder heads, counterflow, 7-9
Cylinder heads, cross-flow, 7-9
Cylinder leakage tester, 19-20
Cylinder measurement, 79-82
Cylinder testing, 154

Dial indicator, 23
Distributor caps, 132, 150

Engine tuning, 143
Engines, four-stroke, 5-6
Exhaust manifolds, 132
Exhaust systems, 140

Feeler gauge, 23
Flame front, 16-17
Fuel filters, 152-153

Gases, intake and exhaust, 15

H-pipes, 141-142
Hand tools, 27

Intake manifolds, 132, 135, 138

Lifters, 69-70, 94, 129
    hydraulic, 130-131
    mechanical, 129-130

Mass airflow sensor (MAF), 136-138
Micrometers, 21-22
Mufflers, 140-141

Nuts, 30

Oil pans, 132
Oil pressure gauge, 20
Oil pumps, 131
Otto Cycle Engine, 13-14

Plastigauge, 23
Point-condenser ignition systems, 155-157
Ports, measuring, 82-83
Pushrods, 69-70, 93-94, 123-125, 129

Quad 4 engine, 6-7

Radiator pressure tester, 25
Ridge reamer, 25-26

Rocker arm guide plates, 123
Rocker arms, 53-54, 93-94, 122-123

Screws, 30
Sensors, testing, 146-150
Split-ball gauge, 22-23
Studs, 31

Tachometer, 23
Telescoping gauges, 22-23
Thermal efficiency, 48
Thermodynamics, 12-13
Throttle body assembly, 138
Timing light, 25
Torque wrench, 20-21
Troubleshooting, 143-146

Valve adjusting clips, 26
Valve covers, 132
Valve faces, resurfacing, 42-44
Valve guide drift, 21
Valve guide inserts, 120
Valve guides, replacing, 42, 74-75, 87-90, 94-95
Valve
    grinds, 45-46, 97
    lapping, 46, 104-105
    masking, 106-110
    overlap, 55-57
    refitting, 118120
    removal and replacement, 71-73
Valve seals, 90-91
Valve seats, 44, 97-104
Valve spring compressor, 25
Valve springs, 72, 75-76, 83-84, 91-92, 94, 96, 120-122
Valve stems, 75, 96
Valves, types of, 85-87
Volumetric efficiency, 48

Washers, 30-31